CHEMISTRY ON THE INTERNET
A STUDENT'S GUIDE
1999 - 2000

THOMAS GARDNER

PRENTICE HALL, Upper Saddle River, NJ 07458
http://www.prenhall.com

Editorial Director: Tim Bozik
Editor in Chief/Science: Paul Corey
Editor in Chief/Development: Carol Trueheart
Acquisitions Editor: John Challice
Special Projects Manager: Barbara A. Murray
Manufacturing Manager: Trudy Pisciotti
Supplement Cover Manager: Paul Gourhan
Supplement Cover Designer: PM Workshop Inc.
Manufacturing Buyer: Ben Smith

© 1999 by Prentice Hall
Upper Saddle River, NJ 07458

All rights reserved. No part of this book may be reproduced, in any form or by any means, electronic or mechanical, including photocopying, recording, or any information storage and retrieval system, without permission in writing from the publisher.

TRADEMARK INFORMATION:
Microsoft Windows and Microsoft Internet Explorer are trademarks of Microsoft Corporation; Macintosh is a trademark of Apple Corporation; NCSA Mosaic is a trademark of the National Center for Supercomputing Applications; Netscape is a trademark of Netscape Communications Corporation; Java is a registered trademark of Sun Microsystems. All other products and trademarks are the property of their respective owners.

The author and publisher of this manual have used their best efforts in preparing this book. The author and publisher make no warranty of any kind, expressed or implied, with regard to these programs or the documentation contained in this book. The author and publisher shall not be liable in any event for incidental or consequential damages in connection with, or arising out of, the furnishing, performance, or use of the programs described in this book.

Printed in the United States of America

10 9 8 7 6 5 4 3 2 1

ISBN 0-13-083977-9

Prentice-Hall International (UK) Limited, London
Prentice-Hall of Australia Pty. Limited, Sydney
Prentice-Hall Canada, Inc., London
Prentice-Hall Hispanoamericana, S.A., Mexico
Prentice-Hall of India Private Limited, New Delhi
Prentice-Hall of Japan, Inc., Tokyo
Simon & Schuster Asia Pte. Ltd., Singapore
Editora Prentice-Hall do Brazil, Ltda., Rio de Janeiro

Preface

The new Information Age is upon us. The Internet, which has existed for about a quarter century, has finally come into its own. How will it affect our lives, and more to the point of this book, how will it change the way we learn chemistry? This book, which introduces you not only to the Internet but also to Prentice Hall's *ChemCentral* World Wide Web site, will give you a greater understanding of the Internet's impact on the ways in which you can access chemical information and learn chemistry.

The first chapter of this book presents an overview of what the Internet is all about and attempts to counteract the popular perception that the Internet is only about web pages and e-mail. In truth, it is far more, and the limits are even today being extended to include new media. This chapter discusses all aspects of the Internet, with an introduction to search techniques for locating just what you want out of that vast sea of information. Chapter Two describes the software and hardware tools necessary to explore the Internet, with special emphasis on the web browser and tools of particular interest to chemists. The third chapter gets you started on your web surfing with some suggestions about how you might use the Internet for your studies in chemistry.

With the fluxional and volatile nature of sites and documents on the Internet, it cannot be guaranteed that all URL addresses listed in this text will remain in place indefinitely. Some may move over time, or disappear entirely. Should you happen to come upon a site that has changed from the listing in this text, please notify me by e-mail at chemnet@hotmail.com. An accurate and timely listing of the references that appear in this book will be maintained at Prentice Hall's *ChemCentral* web site at http://www.prenhall.com/~chem.

Thomas Gardner

Acknowledgments

Some topics in this book have been previously presented in my "Chemistry on the Internet" column for *The Chemical Educator*, an on-line professional journal published by Springer-Verlag at http://journals.springer-ny.com/chedr. All excerpts appear here with their kind permission.

I wish to thank John Challice and Paul Corey of Prentice Hall for providing me an opportunity to share my enthusiasm for this new information medium with all of you. I am also grateful to Cliff LeMaster, editor-in-chief of *The Chemical Educator*, who first gave me a virtual soapbox on which to stand. Last but never least, my appreciation goes out to those students over the years who have taken time out of their busy schedules to show me the latest computer tricks and Internet tips. It's always nice to see the student become the teacher!

Table of Contents

PREFACE iii

ACKNOWLEDGMENTS iv

TABLE OF CONTENTS v

1 INTRODUCTION TO THE INTERNET 1

What is the Internet? 1
   Spreading the Net 2

Types of Internet Communication 3
   Remote Log-in (Telnet) 3
   File Transfer Protocol (FTP) 4
   Electronic Mail (E-Mail) 4
   Usenet News 4
   Gopher 5
   World Wide Web (WWW) 5
   Push Technology 6
   Specialty Applications 6

2 ACCESSING THE INTERNET 7

Hooking into the Net 7
   The Computer 7
   The Connection 8

Know Your Hosts 9
   IP Address and Domain Names 9
   Protocols and URLs 10

Client Software for Navigating the Internet 11
   The Web Browser 11
   Enhancing the Web Browser 14
     Plug-Ins 14
       Chemscape Chime 14
       Shockwave 16
       QuickTime 16
       Real Player 18
       Virtual Reality Modeling Language (VRML) 18
     Java 19
       FormulaOne Java applet 21
     ActiveX 22
     Scripts 23

Using the Web Browser to Perform Other Internet Functions	25
Gopher and FTP	25
E-Mail and Usenet News	25
Telnet	27

Other Useful Software — 27
Chemical Structure Drawing Programs — 27
Molecular Structure Viewers — 27
File Compression/Decompression Software — 30

3 SEARCHING THE INTERNET — 33

General Search Tools — 33
Searching the World Wide Web — 33
Searching Usenet News and Mailing Lists — 34
Searching FTP Sites - *Archie* — 35
Searching Gopher Sites - *Veronica* **and** *Jughead* — 36

Chemistry-Related Searches — 36
Searching by Structure — 36

4 *ChemCentral*: A GUIDED TOUR — 43

GLOSSARY — 49

INDEX — 55

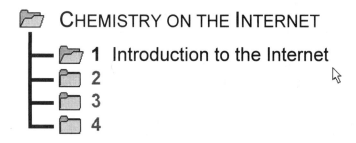

CHEMISTRY ON THE INTERNET

1 Introduction to the Internet
2
3
4

Imagine you had a chemistry textbook that was unlike any other book you owned. On the pages of this book, you had not only text and pictures, but also animations, sounds, 3-D structures that you could rotate at will, calculators built into every page, and the ability to magically call up another book on a specific topic by simply touching a phrase in the first book. Believe it or not, such a "book" exists today for anyone with access to a computer connected to the Internet, using technology that will forever change the traditions of media.

You've undoubtedly heard about the Internet from television or other media. It's hard not to these days. A TV commercial might referred you to a company's World Wide Web site (such as "http://www.something.com") that provides more information on its product, or a long-distance friend might ask you if you have an e-mail address. These are two uses of the Internet, but certainly not the limit of them. The current state of the Internet allows you to hold video phone conversations with someone on the other side of the Earth without long-distance telephone charges, search the catalogues of libraries all over the world and read the great works of literature, obtain a wide variety of software for free, watch live broadcasts of films and concerts, and remotely operate laboratory instrumentation. All these things can be done with just a computer, an Internet connection, and software that can be obtained at little or no cost.

The Internet is the latest step in the evolution of electronic telecommunications technology. Just as the telegraph introduced the ability to send text messages across great distances, telephone and radio permitted voice transmission, and television added video images to the audio, the Internet now combines all aspects of information and multimedia transmission into one generalized technology. What's different about the Internet is that it allows for a higher degree of personal freedom and interactivity. You decide what you want from the Internet, and when you want it. Many Internet sites allow you to make interactive choices on how its information is to be presented. An especially attractive feature for most people is the ability to design and produce your own publications for distribution to a global audience, without the need to go through a commercial broadcaster or publisher. What would you like the world to know?

What is the Internet?

The Internet is an ever-growing collection of several million computers in a vast global network, with each one providing information of some sort, and countless others reading that information. The computers that provide the information are known as *servers*, or *host* computers, and those that read the information (such as the computer you would use to access the Internet) are called *clients*. Each computer connected to the Internet is called a *node*. The Internet is actually an interconnection of smaller regional networks, and it is to one of these that you connect to reach other Internet nodes (just as you must navigate a local residential road system before you can get onto an interstate highway).

Why do we have an Internet? You can thank the U.S. Department of Defense (DOD) for this (or perhaps ultimately, the former Soviet Union). During the Cold War of the 1960s, the DOD wanted to develop a more advanced communications system that would be less vulnerable to disruption in the event of war. Research began on methods of connecting computers into a network matrix that would form an interlaced web. In this way if one communication line were disabled, messages could still be relayed through other computers and then on to their destination. The development of a network that could transfer information between different types of computers was also of primary importance. As a result, the Internet is designed to be largely a platform-independent medium; that is, it can be used as easily by a PC as by a Macintosh or a Unix computer.

The agency of the DOD responsible for developing this communications network was the Advanced Research Projects Agency, or ARPA (which was formed partially in response to the launching of Sputnik, the first artificial Earth satellite, by the Soviet Union in 1957). The first network, ARPANET, went into operation in December of 1969 with only four computers (one each at UCLA, UC Santa Barbara, the Stanford Research Institute, and the University of Utah). Over time, other networks were formed for various purposes (with such names as BITNET, USENET, NSFNET, THEORYNET, and ALOHANET, to name but a few), and these new networks became interconnected into the global matrix that we now know as the Internet.

One of the first things the ARPANET researchers noticed about the network was that, aside from the tasks that it was designed to do, it was also a very convenient medium for relaying routine text messages. These electronically transmitted memoranda quickly became a significant portion of the network traffic. This ancestor of what we now call e-mail foreshadowed much of what the Internet was to become. For decades, information delivered over the Internet took the form of text messages or computer programs; however, in recent years, this has changed greatly with the introduction of multimedia and interactivity.

Spreading the Net

The Internet has come a very long way since those first four computers back in 1969. In recent years, the number of unique host computers connected to the Internet has grown exponentially to over 30 million servers. If the Internet has been around since 1970, why is it we are only now hearing about it in the popular media? Part of this must be attributed to the advent of personal computers in the late '70s and early '80s. More recently, it has been the result of advances in the way information can be presented over the Internet. Where earlier the network traffic consisted primarily of text messages, now all kinds of multimedia are possible. The commercialization of the Internet in the early 1990s, when administration of the networks was passed from the National Science Foundation (NSF) to various corporate entities, also enhanced the Internet's growth and popularity. Certainly, the ability to independently publish multimedia information freely and easily to a planet-wide audience is a major attraction to Internet users.

The Internet's exponential growth has caused some people to liken it to living thing. This analogy is not so far-fetched when you consider that most organisms are composed of countless cells of diverse types, each with a unique function, that symbiotically interact to produce a being that is greater than the sum of its parts. With the Internet, there are millions of servers, each providing unique information to your computer. You are no longer limited to the information that your computer can hold on its hard drive, but rather now have access to millions of times as much data scattered around the world.

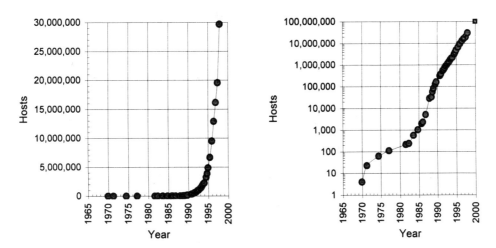

Figure 1-1. *Growth of the Internet. These graphs (shown in linear and logarithmic scales) display the growing number of unique host computers on the Internet since the first ARPANET in December 1969.*

Types of Internet Communication

The manner in which some people perceive the Internet today is reminiscent of a fable from India, "The Blind Man and the Elephant." In this story, a group of blind men happened upon an elephant, and tried to identify it solely by touch. However, because each one of them felt only a specific part of the animal, incomplete mental images were developed that were quite different from the whole. For example, one of them perceived the elephant's tail to be a rope, while another grasped one of the legs and thought it a tree trunk. Some people who use the Internet, or who wish to begin doing so, are in a similar position; they may only use e-mail, or see the Internet as a collection of World Wide Web sites. In order to get the complete picture, let's take a moment to outline the various ways in which the Internet is used. In the next chapter, as we learn the tools for accessing the Internet, we will see specific examples of how each of these methods is used.

Remote Log-in (Telnet)

One of the oldest ways in which the Internet is used is in a process known as ***remote log-in***, or ***telnet***. To *log in* to a computer is to gain access to its contents through some security protocol (such as a user name and password); therefore, using remote log-in is to gain access to someone else's computer. This process is usually used to run a program on that remote computer. These programs commonly include searchable databases of information, and areas for posting and retrieving messages. Your campus library probably allows for telnet connections for the purpose of searching its catalog listings.

There are also many databases on the Internet providing information on chemicals, listings of journal articles, employment opportunities, and other information relevant to scientists. Areas on the Internet that allow for an exchange of messages are known as *bulletin board systems* (BBS). A BBS is a place where people can discuss various topics, either in live one-to-one chat or by posting messages that everyone else can read. Despite being one of the original methods of transferring information over the Internet, telnet is still widely used in situations where the information can be completely represented by simple text.

File Transfer Protocol (FTP)

Aside from running programs residing on a remote computer, another original intent in designing the Internet was to transfer information from one computer to another. This includes not only text files, but also binary files such as programs and images. For this reason, a ***file transfer protocol* (FTP)** was developed. There are a vast number of FTP sites on the Internet that house archives of software and other files of general interest. Descriptions of molecular structures can also be transferred by FTP; this is done by creating text files containing listings of atomic identities, positions in Cartesian coordinates, and information about bond connectivity.

Electronic Mail (E-Mail)

If one can transfer text files from one computer to another, it naturally follows that one can send text messages between computers. This occurred quite readily to early network developers, who found it at least as easy to send text messages to another development site as it was to telephone them. ***Electronic mail* (e-mail)** has become a commonplace means of communication these days, largely because of its lightning-fast speed of delivery relative to traditional mail, and its obvious cost advantage over long-distance telephone calls. An added advantage is the ability to *attach* binary files to e-mail messages, allowing you to send programs, images, or other non-text files to someone else. E-mail can also serve as a forum for discussion, much like a BBS, through what is called a *mailing list*. A message is posted to one of these discussion groups by e-mailing it to a *listserver*, which then sends it on to all subscribed members of the list.

Usenet News

For the ultimate in discussion group participation, one turns to ***Usenet News***. This is not "news" in the sense of journalistic media, but rather an extremely large BBS that is comprised of tens of thousands of specific discussion areas called newsgroups. Unlike mailing lists, newsgroups are not limited to subscribers, but rather are open to anyone wishing to post or read messages. In contrast to BBS discussion forums, many Usenet newsgroups allow the posting of binary files, thus removing the limitation of text-only postings. Some newsgroups have been in existence for many years, and as a result, recurrent topics and questions tend to appear. For this reason, a regular posting to many newsgroups is the FAQ, or Frequently Asked Questions document, prepared by one or more regularly active participants of the newsgroup. The FAQ has become such a common type of document that it appears in other contexts frequently on the Internet.

Gopher

A great leap forward in the usability of the Internet came in December of 1991, when computer researchers led by Mark MaCahill at the University of Minnesota developed a menu-driven system for presenting sets of interrelated documents over the Internet known as ***gopher*** (a homonym of "go for;" also, Minnesota's school mascot happens to be a gopher). Gopher offers directories of folders and documents (either text or binary), reminiscent of the directory trees found in most computer operating systems. Although FTP sites also use directory trees, gopher sites can also contain some menu items that allow you to jump from the site you are currently visiting to an entirely new site. Prior to gopher, navigating the Internet meant logging into one telnet or FTP site, and when you wanted to change to a new site, logging out and then logging in at the new site; with gopher, however, jumping from one site to another typically requires no such login procedures (except at protected sites). This improvement in the ease with which the Internet could be navigated heralded a period of rapid growth in its use and popularity.

World Wide Web (WWW)

The latest and perhaps greatest step in the evolution of the Internet came one year later, when Tim Berners-Lee of CERN (the high-energy physics research institute in Geneva, Switzerland) devised a system whereby the "menu list" structure of gopher was modified into more natural form called *hypertext*. In hypertext, connections to documents or other sites were mixed into the text of a document, such that key words or phrases in the body of the text acted as *links* that would lead out of the document. The second advantage to hypertext is that it allowed for the formatting of a document; instead of a plain-text document, images and other objects could be included and arranged within the document, and the text and layout of the document could be formatted. This makes hypertext documents ideal for viewing with software based on graphical user interfaces (GUI) such as Windows or Macintosh. In addition to key words and phrases of a document, the images in a document can also be made to act as links. These links not only can be to other hypertext documents, but also to gopher, FTP, Usenet, and telnet sites, and even to e-mail addresses. The formatting of a hypertext document and the definition of its links are achieved through what is called *hypertext markup language* (HTML). This consists of angle-bracketed code letters, or *tags*, placed throughout the text (such as `<B>` and `</B>` for the beginning and end of the section of text to be displayed in a bold font). The collection of hypertext documents on the Internet form a subset of the Internet known as the ***World Wide Web*** **(WWW)**, or simply "the web" for short, and the documents themselves are referred to as *web pages*. Web pages that specifically act as a main or front page for other pages in a web site, or web pages on a particular person or subject, are called *home pages*. The introduction of HTML documents to the Internet allowed for the transition from plain text to attractively and creatively designed documents comparable to what one might see in print media, but with the added features of animation, sound, and interactivity.

Push Technology

A recent spin-off of hypertext is ***push technology***, which is designed to deliver a packet of web pages containing information that is likely to be frequently updated. These might include, for example, weather reports, stock market updates, news bulletins, or periodicals and journals. In software designed to receive such "pushed" information, all pre-selected pages can be downloaded with a single command (or at a predetermined interval) and viewed at the reader's leisure.

Specialty Applications

As the Internet continues to evolve, specialized applications appear at an increasing rate. Chief among these are the ***chat*** applications, which allow two or more users to communicate by voice, and in some cases with video. Advanced chat applications allow the exchange of other live images during the chat. Other applications retrieve information from specialized servers, such as: time servers, where you can set your computer's clock against an atomic clock; weather servers, where you can get up-to-the-minute forecasts; and financial servers, where you can get current stock market information. There are also applications more directly related to Internet functions, such as ***finger***, which can give you information about another Internet user.

In the next chapter, we shall look at how to connect to the Internet and where to obtain the tools necessary for exploring the Internet and how to use them. We shall also delve more deeply into the various facets of the Internet and how they apply to your study of chemistry.

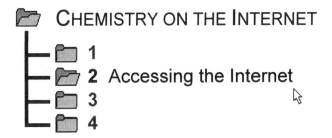

Hooking into the Net

Now that you've seen what the Internet is all about, the question that is probably foremost on your mind is, "How do I get connected to all of this?" Minimally, of course, you'll need a computer and a connection.

The Computer

The type of computer you need depends on how you want to use the Internet. If you require a text-only connection (for example, if all you want out of the Internet is to use e-mail), almost any computer should be fine. Software designed to utilize a graphic user interface (GUI) platform (such as on an Apple Macintosh or on a Microsoft Windows-based system), that uses icons and other graphics to create a point-and-click interface, will require something more robust. Modern GUI platforms require, at the very least, a processor type such as a 486 (for a PC) or 68040 (for a Macintosh), and a minimum of 16 megabytes (MB) of random access memory (RAM, the built-in memory chips inside your computer). GUI-based software for reading information from the Internet is referred to as *client software* (as opposed to the software that provides the information, known as *server software*). You may quickly find that a 486 or 68040 processor and 16 MB of memory are not enough for most of the modern and highly sophisticated WWW client software discussed in this chapter, and that having a Pentium or PowerMac processor with at least 32 MB of RAM will speed up your work and result in fewer system "crashes."

In order to connect to the Internet, your computer will need to have some sort of communications hardware installed inside of it. The type of hardware required depends on how you plan to connect to the Internet: if you'll be connecting through a telephone line (such as from an off-campus residence), you will need a *modem*; if, on the other hand, you have a "dedicated" (full-time) communications line to a *local area network* (LAN), you will need a circuit board for networking (typically an *ethernet card*). Modems and network cards are categorized by the rate at which they transfer information, which is measured in bits per second (bps). Modern modems for use with standard telephone lines operate at speeds on the order of thousands of bps, or kbps, with a top speed of 56 kbps (or simply "56K"). This top speed is limited by the analog nature of the phone line; digital (ISDN) phone lines can transfer information at over twice this speed using an ISDN modem (which, as you might expect, is generally more expensive setup than a standard analog modem and phone line). Connections made through a network are much faster than those made through a telephone line. Network connections are typically used for computers on campuses and in businesses and other institutional buildings. Users at home who wish to overcome the slower speed of modems have the option of establishing a network connection through their cable TV service, which can provide connection speeds at about 500K. While this is not available in all areas yet, it is rapidly becoming more commonplace.

The Connection

It is fortunate that connecting to the Internet is far easier than it has ever been in the past. The actual connection is usually made in one of two ways: either through a public-access terminal, or through your own computer and an *Internet service provider* (ISP). Most campuses make computers available to students for use in their studies at little or no cost, and these are often connected to a campus network, which in turn is connected to the Internet. Libraries also often have public-access terminals connected to the Internet, as do *Internet cafes* (places for dining and socializing while you navigate the Internet) and office service/photocopy shops. An advantage to using public access terminals is that they are already set up with the necessary basic software for navigating the Internet, and a connection has already been configured; however, if you require specialized software, you may be out of luck. Even if you have access to campus terminal, you may be well advised to make an arrangement with an ISP through which you can access the Internet with your own computer. This arrangement has several advantages: because you're using your own computer, you can configure it however you like and load it with the software that you especially need; with an ISP, you can also remain connected to the Internet in the same way whether you are on or off campus (especially useful if your campus server does not allow dial-up access); in addition, you may find that you get more and better services through a commercial ISP than through your campus server.

Selecting a good ISP requires some careful shopping. Advertisements for ISPs can be found in the telephone directory (generally listed under "Computers," "Internet," or "Network"), or in computing magazines (a better source for finding national services). ISPs can be local to a particular city, regional to a state or telephone company service area, national, or global. This is important to keep in mind, for example, if you go to school in one part of the country and vacation with family in another, and want to stay connected to the same ISP the entire time. (This is helpful to people who are trying to reach you by e-mail, as changing ISPs means changing your e-mail address.) The larger ISPs will have local dial-up access numbers in most major cities; if you live in a smaller town, you should make sure that the dial-up number that you use does not incur long distance charges. A good ISP will provide you with the necessary software for your system to connect to the ISP's modem and navigate the Internet, and will provide you with clear and detailed instructions on how to configure your computer for the connection. You should also consider the services that the ISP has to offer: most will give you an e-mail account, a certain amount of space on their server for your web pages, and access to a large number of Usenet newsgroups. The cost of the service is, of course, a primary consideration, and should be weighed against the services offered. Some services charge for access by the hour, while many have adopted a monthly flat rate fee; still others offer a choice between the two rate structures.

When shopping for dial-up access to an ISP, try to find one that provides either *serial line Internet protocol* (SLIP), *compressed SLIP* (CSLIP), or *point-to-point protocol* (PPP) connections. These are the telephone connection types that allow for the simultaneous transfer of text, images, and other files in GUI client programs such as web browsers; they also allow you to use more than one client program at once. PPP is somewhat more reliable in transferring data than CSLIP or SLIP. Without one of these protocol types, you will almost certainly be constrained to a text-only connection. You should also find out the maximum modem speed that the ISP supports; it doesn't pay to have a fast, expensive modem if your ISP cannot receive data at that rate.

Know Your Hosts

IP Addresses and Domain Names

Once you become connected to the Internet, it then falls upon you to know how to find your way around. With millions of servers on the Internet, how do you tell one from another? Just as telephones have unique phone numbers and buildings have unique street addresses, Internet nodes (both servers and clients) have unique addresses, known as *Internet protocol (IP) addresses*. Each IP address consists of a set of four numbers, each between `0` and `255`, separated by decimal points. When your computer dials into an ISP, their server automatically assigns a *dynamic* IP address to your client computer, and a *router* then helps you to connect to other IP addresses upon your request. Campus computers connected through a local network usually have permanent, or *static*, IP addresses assigned to them.

Would you like to be using twelve-digit phone numbers on a regular basis? Probably not. Because long numbers can be tricky to remember, the *domain name system* (DNS) was adopted. A *domain name* is a unique name assigned at the option of a host computer's system administrator and registered through an organization called *InterNIC*; it consists of a minimum of two alphanumeric words, again separated by decimal points. While every node on the Internet has an IP address, not every one needs to have a domain name as well. The last part of the name (the *domain* itself) gives an indication of the general location or type of the host computer. Commonly used domains include:

`.com`	commercial	`.mil`	military
`.edu`	educational	`.org`	non-profit organization
`.gov`	government	`.net`	other networks

A domain might also be a two-letter code symbolizing the country in which the host computer resides; for example, Canada's domain is `ca`, the United Kingdom's is `uk`, Germany's is `de` (for Deutschland), and so on. Typically, hosts located in the United States do not carry the `us` domain, except for state and local government organizations, where the two-letter state code precedes `us`. Other parts of the domain name are designed to help you remember details about the host. For example, a name might be prefixed with `www` for a World Wide Web site, or `ftp` for a file repository (these terms will be defined more fully in the next section). Libraries often put `lib` in their domain name as well.

Let's look at some examples of IP addresses and domain names. Either can be used when accessing a remote computer, although it is customary (and convenient) to use a domain name when one is available.

Host	IP Address	Domain Name
Prentice Hall	192.251.132.3	`prenhall.com`
American Chemical Society (gopher site)	134.243.201.30	`gopher.acs.org`
Tennessee State University	198.146.80.11	`acad.tnstate.edu`
U.S. Dept. of the Navy (ftp site)	138.147.50.5	`ftp.navy.mil`
The White House	198.137.240.92	`www.whitehouse.gov`
Ohio Public Library	206.244.97.141	`www.oplin.lib.oh.us`

Protocols and URLs

In addition to identifying the address of the Internet server to which you wish to connect, it is usually also necessary to identify the type of information transfer you plan to make. For example, are you calling upon a web site, a gopher site, an FTP site, or some other type of site? These different types are referred to as a *protocols*; the most common protocol types are listed in the table below with their corresponding prefixes.

Prefix	Description
`http://`	*hypertext transfer protocol - Web site*
`gopher://`	*gopher site*
`news:`	*Usenet newsgroup* (Notice a double-slash is not used here.)
`ftp://`	*file transfer protocol - FTP site*
`telnet://`	*telnet address*
`mailto:`	*e-mail sender* (Again, a double-slash is not used here.)
`file:///`	*local file* (Notice a triple-slash used in this protocol type.)

The protocol designation followed by the host server address (*i.e.*, a domain name or IP address), directory information and file name constitutes what is known as a *uniform resource locator* (URL). It is this command line that instructs most Internet client software in how to locate and connect to a destination. All parts of the URL are separated by slashes, and no spaces are included; also, the type is case sensitive. The syntax of a typical URL is shown below, in this case for the web site that is home to Prentice Hall's *ChemCentral*.

```
http:// www.prenhall.com /~chem/ index.html
```

| Protocol Type | Host Address | Directory Path | Document Name |

Client Software for Navigating the Internet

Software for every aspect of Internet communication is available for Macintosh, Windows, and Unix computers, and much of it is free, especially to educational and nonprofit users. By far the most useful client program is the *web browser*. Just as gopher allows FTP and telnet operations, the WWW is compatible with almost all other major Internet operations, and a good web browser can meet nearly every common need.

The Web Browser

A web browser is a software application that is designed to present web pages, as well as less-sophisticated documents, in a formatted manner. It also provides an interface for the input of information and the output of processed results. In a sense, a web browser is like an operating system in microcosm, in that it can run smaller programs within itself, such as animation and audio players, molecular structure visualizers, spreadsheets, and a wide variety of other specialized applications.

The first browser to surf the web was *Mosaic*, introduced in 1993 by Marc Andreessen at the National Center for Supercomputer Applications (NCSA) at the University of Illinois at Urbana-Champaign. This software, which was the first to show both text and images on the same document page, met with instant popularity and success. Shortly after that, Andreessen started his own company, Netscape Communications, and designed a new browser called *Netscape Navigator*. Many other browsers have been developed in the interim, including a modern version of Mosaic and *Internet Explorer* from Microsoft, which loosely parallel the features found in Netscape Navigator. Navigator, however, remains the most popular browser in use today and is the one used for the majority of the examples in this text.

At the time of this writing, the most up-to-date release version of Netscape's web browser is Navigator 4.5, and Microsoft's Internet Explorer is at version 4.01. The Netscape Navigator browser is bundled as part of the Communicator package of Internet-related applications; Internet Explorer is similarly bundled with software programs designed to enhance the Windows operating system. Both standalone browsers are available for free and may be downloaded from the Internet at various locations; at least one type of browser is also typically included with new computers and on operating system CD-ROMs. A browser is also usually a part of the software package provided to a new customer by an ISP.

The web browser may be one of the most user-friendly applications in existence today. Navigation to a web page can be accomplished by typing a URL address into the Location Toolbar, by selecting a preset Bookmark, or by selecting a link in a web page. Figure 2-1 describes a typical web page in a browser window, in this case Prentice Hall's *ChemCentral* web page, as it appears in Netscape Navigator. The links in this web page can be found in many places. As is typically the case, links can be made through text, as seen in the blue (and in some browsers, underlined) text of the disclaimer. Images can also serve as links, such as the title image, or the circular menu in the center of the page (which is actually an *image map* that contains multiple links through one image). A link can also be activated through the use of buttons and forms, such as the *radio buttons* on the left-hand side of this web page.

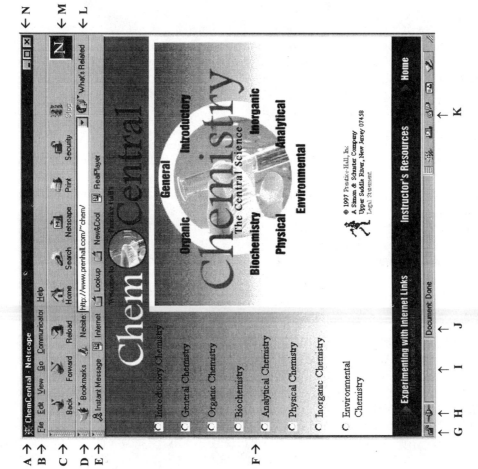

Figure 2-1. *A typical web browser window (in this case Netscape Navigator), showing Prentice Hall's ChemCentral web site. The individual functions are described in the table below.*

Key	Item	Description
A	Title Bar	This line tells you the title of the web page you are viewing, and what browser you are using. The icon on the left controls the window.
B	Windows Menu Bar	This is a standard Windows menu bar.
C	Navigation Toolbar	These buttons are for navigation through the *history list* (sites you've visited in the current session), and for other functions such as reloading or printing the page and searching the page for specific text.
D	Location Toolbar	This toolbar has two functions: the bookmark button allows you to save and recall the addresses of pages; the locator box is where you type the address of the site you wish to visit, and it also displays the address of the current window.
E	Personal Toolbar	These buttons connect you to Netscape web pages with various helpful features, or start other programs to perform Internet tasks.
F	The Web Page	The bulk of the window area is allocated to the display of the web page.
G	Security Status	This tells you if a page is using a secure connection.
H	Connection Status	
I	Activity Indicator	This becomes animated while a page is loading.
J	Page Status	This tells you the progress of page loading, the location of links when you hold the mouse cursor over a link, and other special messages.
K	Communications & Composition Icons	Clicking on these icons will open e-mail, Usenet news, and web page composition applications.
L	What's Related	This tool can be used to search for pages that might be related to the one currently being viewed.
M	Browser Icon	This becomes animated when a page is loading. Clicking on this icon takes you to that browser's home page.
N	Windows Controls	These are standard Windows control toggle buttons for minimizing, maximizing, and closing a window.

Enhancing the Web Browser

As was stated earlier, a web browser is like a miniature operating system, in that it can run smaller programs within itself. The net result of this is that web pages become greatly enhanced through the use of these specialized applications. This enhancement can be done in one of four ways: through the use of *plug-ins*, *Java*, *ActiveX*, or *scripts*.

Plug-Ins

A plug-in is a program that the user installs once, and afterward it remains resident on the system for use by the browser. Plug-ins may be downloaded either from the web site of the company that authored them, or from central repositories such as Netscape's plug-in page,[1] *BrowserWatch*,[2] or *TUCOWS*.[3] There are several plug-ins that you, as a chemistry student, will find useful, including *Chemscape Chime*, *Shockwave*, *Real Player*, a *Virtual Reality Modeling Language (VRML)* viewer, and *QuickTime*.

Chemscape Chime

Chemscape Chime, a browser plug-in published by MDL[4] for both Netscape Navigator and Microsoft Internet Explorer, is used to visualize molecular structures in 3-D within a web page. Molecules can be placed in-line with the text and graphics of a web page and manipulated at will. The user is given a selection of display modes for the molecule, including wireframe, ball-and-stick, and space-filling. A molecule can be rotated by dragging a clicked mouse over it. Molecules that have animation information in their structure files can even be shown to move (i.e., bond vibration and rotations). Chime also has a rich scripting language associated with it that controls many advanced display features. It evolved from *RasMol*, a program that has mostly the same functionality as Chime. However, RasMol is not a plug-in and runs separately from the browser.

A Chime 2.0 installation program[5] can be downloaded from the MDL web site. It will place Chime on your system and instruct your browser to recognize and display molecular structure files when it encounters them on the web. Once you have installed Chime, you can test[6] it at the MDL site to see some examples of how molecules can be displayed. A pop-up floating menu of display commands (Figure 2-2) can be accessed by pressing the right mouse button over the molecule (or for Macintosh users, pressing the mouse button until the menu appears).

After you've confirmed that Chime is properly installed, you can try out a few sites. MDL offers a list of sites that use Chime images.[7] Check out John Nash's *Structure and Isomerism of Coordination Compounds*[8] (Figure 2-3) and *Valence Shell Electron Pair*

---

[1] http://home.netscape.com/plugins/
[2] http://browserwatch.iworld.com/plug-in.html
[3] http://idirect.tucows.com/
[4] http://www.mdli.com/
[5] http://www.mdli.com/download/chimedown.html
[6] http://www.mdli.com/support/chime/sample.htm
[7] http://www.mdli.com/support/chime/cool.htm
[8] http://www.chem.purdue.edu/courses/chm116/test/chime/

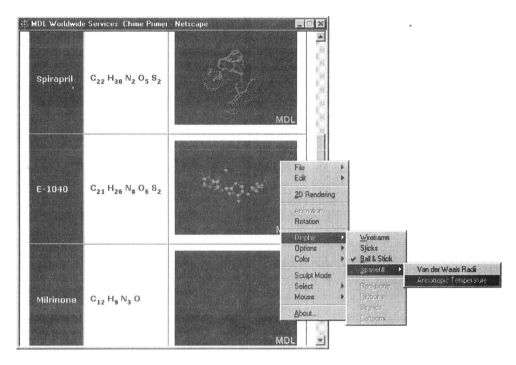

Figure 2-2. *Chime displays from the MDL web site. A pop-up menu controls the display properties.*

Figure 2-3. *A sample page from John Nash's* **Structure and Isomerism of Coordination Compounds.** *(Notice the use of frames in this page; they allow multiple sub-pages to appear in one window.)*

Repulsion (VSEPR) [9] pages from Purdue University, as well as Mark Winter's *VSEPR*[10] page from the University of Sheffield in the U.K.; these sites use Chime to illustrate the fundamental concepts of molecular geometry. Over 1100 compounds are on the *Molecular Models*[11] page at Okanagan University College in British Columbia, Canada. Finally, try the National Cancer Institute's[12] database of molecular structures (there are over 400,000 molecules there from which to choose). Of course, be sure to make use of molecular structure files at *ChemCentral*,[13] which offers 3-D representations of the molecules shown in the various Prentice Hall chemistry textbooks.

Shockwave

Macromedia's Shockwave[14] software offers a way of presenting interactive audio-video animations over the web. Plug-ins that allow you to play Shockwave files may be downloaded for free from the Macromedia web site. Shockwave animations are an excellent way to simulate laboratory experiments, as can be seen on the Explore Science web site.[15] Figure 2-4 walks you through one such experiment, showing the interactive steps in which a user drags an object from instrument to instrument in order to make measurements of the object's properties, and then finally drags it to a pail of liquid to test a prediction.

In Prentice Hall's *ChemCentral*, listings of web resources can be found for various textbooks. Those sites in the listings that utilize Shockwave technology are identified as such.

QuickTime

Apple QuickTime (QT) [16] is a format for playing a regular (non-interactive) movie. QuickTime technology has been used for many years, especially in CD-ROM animations. The QT plug-in is a built-in feature of newer browsers, and a separate plug-in is available from Apple for older browsers. Having the QT plug-in allows you to view movies of chemical phenomena and laboratory procedures, among other films.

There is also an interactive version of QT called QuickTime Virtual Reality (QTVR) . With a QTVR file, one can obtain a panoramic view of a scene at all angles. This will have future applications over the Internet, especially in viewing large biological molecules from an interior perspective.

Again, Prentice Hall's *ChemCentral*, web resource listings are marked to show those sites that use QT technology.

---

[9] http://www.chem.purdue.edu/courses/chm116/test/vsepr/
[10] http://www.shef.ac.uk/~chem/vsepr/
[11] http://www.sci.ouc.bc.ca/chem/molecule/molecule.html
[12] http://epnws1.ncifcrf.gov:2345/dis3d/3Ddatabase/dis3d.html
[13] http://www.prenhall.com/~chem/
[14] http://www.macromedia.com/
[15] http://www.explorescience.com/
[16] http://www.apple.com/quicktime/

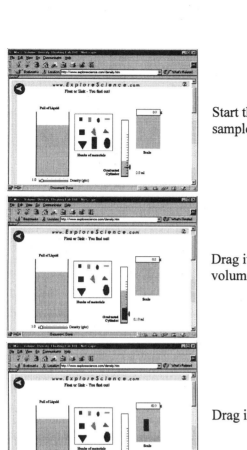 Start this experiment by selecting an object from the sample tray…

Drag it over to the graduated cylinder to measure its volume…

Drag it over to the scale to weigh it…

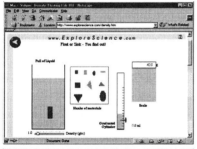 Set the density of the liquid in the pail, predict whether the density of the object will cause it to float or sink, and drag the object over to the beaker…

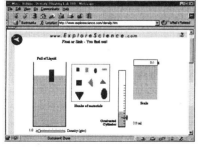 And watch the object (in this case) float to the top!

Figure 2-4. *A Shockwave-based experiment to predict and observe the buoyancy of various objects.*

Real Player

A disadvantage of QT technology is that a file must be downloaded before the animation can be played, and QT files are often several megabytes in size. An alternative to this process is to use *streaming video* technology, in which a video begins to play while it is downloading. (Obviously, one should have a fairly fast connection for this technology to be most effective.)

Real[17] publishes a plug-in and a separate program to play streaming audio and video called Real Player. This plug-in makes it possible to watch live and pre-recorded broadcasts of lectures and other events through a web page (many television and radio stations worldwide have begun to broadcast in this format for web audiences). The separate program also connects to the Internet without a browser to serve the same purpose; however, in using a plug-in, one can place the streaming signal in amongst surrounding text and images to provide more context and spice up a web page. (Another separate program for receiving streaming broadcasts is Microsoft's newest version of Windows Media Player.[18])

Application of streaming audio-video technology to the area of chemical education is still very new; however, in the meantime, the Real Player plug-in is fun and useful to have.

Virtual Reality Modeling Language (VRML)

While Chime can do a remarkably good job of displaying molecular structures, that is just about all that it is designed to do, and there are a great many more 3-D shapes that are relevant to the chemical sciences. For this reason, another useful plug-in to have is one that can display *virtual reality modeling language* (VRML)[19] files. Virtual reality in a browser window is another way to display and manipulate 3-D images, in the same way as Chime only more generally so. A user can rotate an image to view it from all angles, or even from an interior perspective. Netscape Navigator comes with Cosmo Player,[20] a virtual reality player plug-in that can also be obtained separately.

There are many examples of 3-D models in chemistry. Most of these can be found through the *VRML in Chemistry*[21] page, including biomolecules, macromolecules, and crystal lattices. This site also offers some very nice VR renditions of the atomic orbitals (see, for example, Figure 2-5).[22] Web resource listings in *ChemCentral* that offer VRML files are so marked.

---

[17] http://www.real.com/
[18] http://www.microsoft.com/windows/mediaplayer/download/default.asp
[19] a) http://www.netscape.com/eng/live3d/intro_vrml.html
b) http://www.netscape.com/eng/live3d/howto/vrml_primer_index.html
[20] a) http://www.cosmo.sgi.com/
b) http://www.netscape.com/intel/build/products/cosmoplayer2/index.html
[21] http://ws05.pc.chemie.th-darmstadt.de/vrml/
[22] http://ws05.pc.chemie.th-darmstadt.de/vrml/wave.html

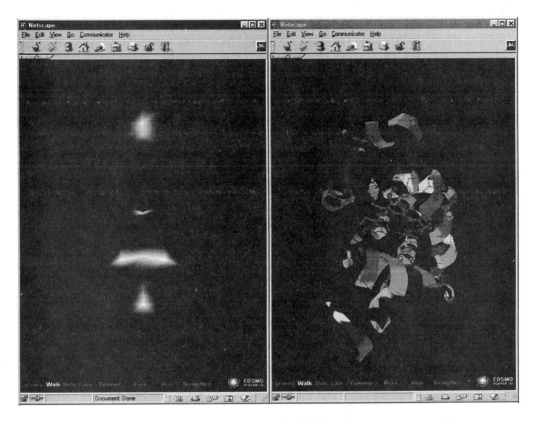

Figure 2-5. *Two examples of VRML images: a d_{z^2} orbital (left), and a representation of a heme molecule. The images can be rotated by the user.*

Java

The Java programming language, developed by Sun,[23] is similar in many ways to the popular object-oriented language C++ used to create many of today's most popular applications. The primary difference, however, is that Java is a language designed to provide programs that can run on any type of computer platform (and even computer controlled appliances and equipment). Java gains this platform-independence through the use of what is called a *virtual machine*: a program written in a language that the computer's operating system understands, which in turn interprets the Java programs and translates them into understandable instructions for the computer. This approach helps remove the long-standing "Tower of Babel" problem of divergent computer languages; indeed, there is considerable effort being applied towards the development of computers that use Java as their primary operating system language.

A web browser (being a mini-operating system) is able to run Java applications (known as *applets*, for "little applications"). When a browser encounters a Java applet, it automatically downloads it and stores it for use in the current user session. Netscape Navigator and Microsoft Internet Explorer both run Java applets; however, Microsoft

---

[23] http://www.java.sun.com/

uses a version of Java that is subtly different from that developed by Sun, and not every applet runs equally well on both browsers (one good argument for keeping a copy of both browsers on one's computer).

One of the first Java applets was a method of displaying molecular structures[24] on a web page (Figure 2-6). These molecules could be rotated in the browser window with a mouse (much like Chime, which didn't exist at the time). The molecules were composed of simple spheres, without bond information or options for alternate display modes. Since then, an ever-increasing number of Java applets have been developed for chemistry. A more sophisticated molecule-viewer applet, ChemSymphony,[25] has since been developed which rivals the capabilities of Chime. Several laboratory experiments have also been simulated using Java applets.

A current and comprehensive listing of Java applets for chemistry and other uses can be found at the *developer.com* web site (formerly known as *Gamelan*).[26] *ChemCentral's* web resource listings also identify those sites that make use of Java applets.

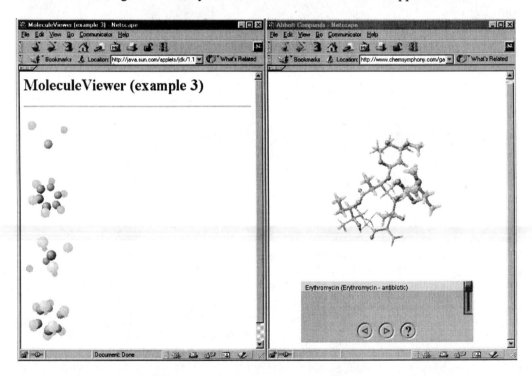

Figure 2-6. *Two Java applets: MoleculeViewer (left), one of the earliest Java web applets, and ChemSymphony (right), a more modern structure viewing applet.*

---

[24] a) http://java.sun.com/applets/jdk/1.1/demo/MoleculeViewer/example1.html
b) http://java.sun.com/applets/jdk/1.1/demo/MoleculeViewer/example2.html
c) http://java.sun.com/applets/jdk/1.1/demo/MoleculeViewer/example3.html
[25] http://www.chemsymphony.com/
[26] http://www.developer.com/directories/pages/dir.java.educational.chemistry.html

FormulaOne Java applet

One important Java applet to know about is FormulaOne, published by Visual Components.[27] This applet provides Internet users a web-based spreadsheet that has many of the same capabilities of the popular Microsoft Excel application, including the ability to create charts that are linked to spreadsheet data. Given how often chemists use mathematics in their work, being able to express this in spreadsheet form over the web is invaluable. An example of this is shown in Figure 2-7, which shows how FormulaOne can allow chemists to simulate titration curves for various acid and base analytes, and thereby identify them.

The FormulaOne applet requires the most recent version of Netscape Navigator running under Windows 95 or 98 in order to function optimally. It should also be noted that an older version of FormulaOne is still available as a plug-in[28] and can be used to view some types of FormulaOne spreadsheet files on older browsers. Even though the plug-in version is older, it does allow one to save files locally, which the Java version does not readily do (although one can use Sun's *Java Development Kit* (JDK) to create a local copy).

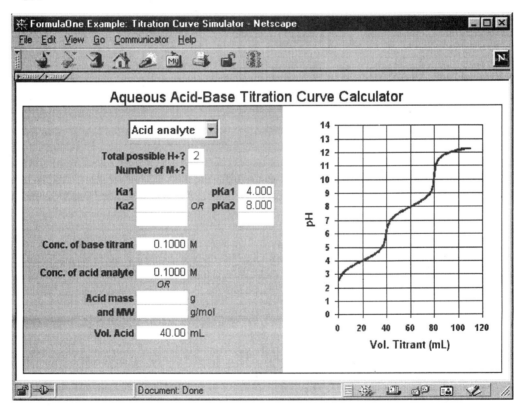

Figure 2-7. *A FormulaOne-based spreadsheet for predicting the shape of acid-base titration curves.*

---

[27] a) http://www.f1j.com/
    b) http://www.visualcomp.com/
[28] http://www.visualcomp.com/products/fonet/default.htm

ActiveX

A Microsoft alternative to the Java applet is the *ActiveX component* (or *ActiveX control*).[29] ActiveX offers something of a compromise between plug-ins and Java: like Java, ActiveX components are downloaded on demand; however, once on the system, it is stored indefinitely like a plug-in. Because of its potentially long-term residence on a user's system, the user is presented with a security *certificate* and asked permission to allow the component to be stored on the system (Figure 2-8). The reason behind this is that malicious programs can be passed to a users system through Java and ActiveX programs (which is also why it is not common for these programs to be able to save files to a users disk drive).

A limitation to the widespread use of ActiveX is that it is designed for use only with Microsoft Internet Explorer; however, this is likely to change in the future, and third-party developers are creating plug-ins to allow Netscape to use ActiveX components. There are also no widespread ActiveX components specifically for chemistry at the time of this writing.[30] Some more general components, such as those for creating graphs,[31] will find uses in chemistry, however. There is also an ActiveX version of FormulaOne spreadsheet program under development.[32]

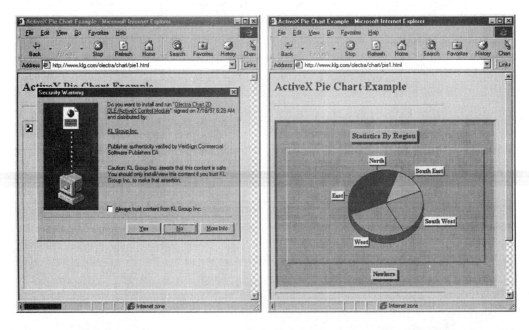

Figure 2-8. *Before an ActiveX component can be loaded onto a system, a security certificate is usually presented (left) to permit the user to decide whether or not to accept the component before the page is completed (right).*

---

[29] http://www.microsoft.com/com/activex.asp
[30] http://jars.developer.com/listing-ActiveX.html
[31] http://www.developer.com/directories/pages/dir.activex.graphics.charts.html
[32] http://www.visualcomp.com/products/fo/default.htm

Scripts

Another way in which web pages can be enhanced is through the use of scripts. Scripts are programming instructions entered directly into the HTML code of a web page. They can be used for a variety of purposes, including controlling the format of the web page, validating and processing the input from a *form* (Figure 2-9), and performing mathematical calculations (Figure 2-10). A script can be *client-side*, being transferred as part of a web page's HTML code, or *server-side*, residing on the server and being called when needed by the client.

There are three major client-side HTML scripting languages currently in use: *JavaScript*, *JScript*, and *VBScript*. JavaScript was the first and today perhaps the most widely used web page scripting language, developed by Netscape and Sun as a "lite" alternative to the Java language. (Because of this, JavaScript can be used to control Java applets through a browser feature called *LiveConnect*.) JavaScript works with both Netscape Navigator and Microsoft Internet Explorer, but not always equally well, as there are subtle differences in the language "vocabulary" accepted by each browser. There are also differences across platforms; a script that works under a Windows browser might not work with a Macintosh browser. JScript is a scripting language created by Microsoft that is similar to JavaScript. Microsoft has also developed VBScript, based on the Virtual Basic programming language. As one might expect, JScript and VBScript are designed to work more often with Internet Explorer than with Navigator; VBScript is especially intended to interact with and control ActiveX components. Like Java, a script is only run when the page is being viewed, and does not stay resident on the user's system. It is also inhibited from writing data to the client's disk drive. Server-side scripts can also be written in these three languages, as well as the computer language *PERL*.

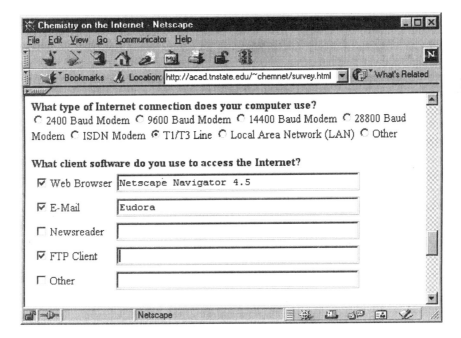

Figure 2-9. *A web page that uses a form for the input of data.*

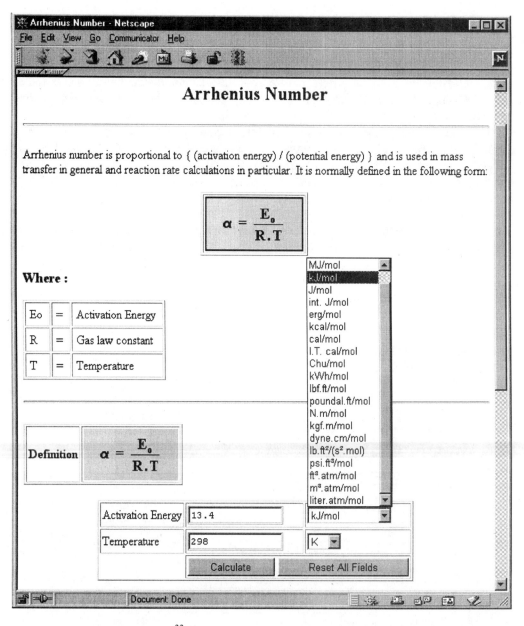

Figure 2-10. *A typical web page[33] that uses a form to allow the user to enter data for a calculation of the Arrhenius number. The data is entered through the text boxes and menu lists. After the* **Calculate** *button is pressed, a new page appears with the result of the calculation.*

---

[33] http://www.processassociates.com/process/dimen/dn_arr.htm

Using the Web Browser to Perform Other Internet Functions

Despite the obvious intent of designing the web browser to show web pages, browsers can also serve many other Internet functions. While there are many other client programs for these functions, you may find your needs are fully met by your browser.

Gopher and FTP

Although the WWW has supplanted most gopher sites in the world, gopher has the advantage of not being bogged down with large images and other features that slow down the load time for a page. For this reason, many documents still exist in a gopher form. Gopher sites can be called using a web browser; the URL must begin with gopher://. Figure 2-11 shows navigation through a typical gopher site, the chemistry resource listing at the University of California at Berkeley's *InfoLib* gopher site.[34]

Throughout this chapter, mention has been made of obtaining software from the Internet, and this would most obviously be done by FTP. Fortunately, it is convenient to perform FTP through a web browser. FTP sites can be opened using the ftp:// protocol in the URL, and navigated in a fashion similar to that used for gopher sites. Files can be obtained even more easily, however, through links in web pages. A web page link might be made to a Windows program file (with a .exe file extension, for example); when one clicks on this link, the browser will merely ask the user where to store the file, and then proceed to download it.

E-Mail and Usenet News

Modern-day browsers incorporate applications for sending and receiving e-mail and Usenet news messages, and these can be readily configured with the consultation of the ISP. One can also use the web to send and receive e-mail and Usenet news.

Several companies[35] offer a free web-based e-mail service, in which a user is given password access to an account in which they use forms to compose e-mail messages, and retrieve messages simply by selecting links. Most of these services can also retrieve messages from another e-mail account (with a password, of course) and are ideal for a person who travels, or even needs to check campus e-mail from off-campus.

Usenet news can also be accessed through the web, primarily through the web site of the *DejaNews Research Service*.[36] This service can be used to read archived Usenet news postings, and can be personalized in an area called *My DejaNews*[37] that sets up a news-reader specific to those newsgroups in which you are interested. (In using My Deja News, the web site leaves a tiny amount of data on your computer to which it refers on return visits, and as such, this service is best used not on a public access terminal but rather on one's own computer.) Access to science-specific newsgroups[38] is also available at the site of the free web-based newsmagazine *Science Daily*.[39]

---

[34] gopher://infolib.lib.berkeley.edu/
[35] http://dir.yahoo.com/Business_and_Economy/Companies/Internet_Services/Email_Providers/Free_Email/
[36] http://www.dejanews.com/
[37] http://www.dejanews.com/rg_reg.xp
[38] http://www.sciencedaily.com/discuss/newsgroups.htm
[39] http://www.sciencedaily.com/

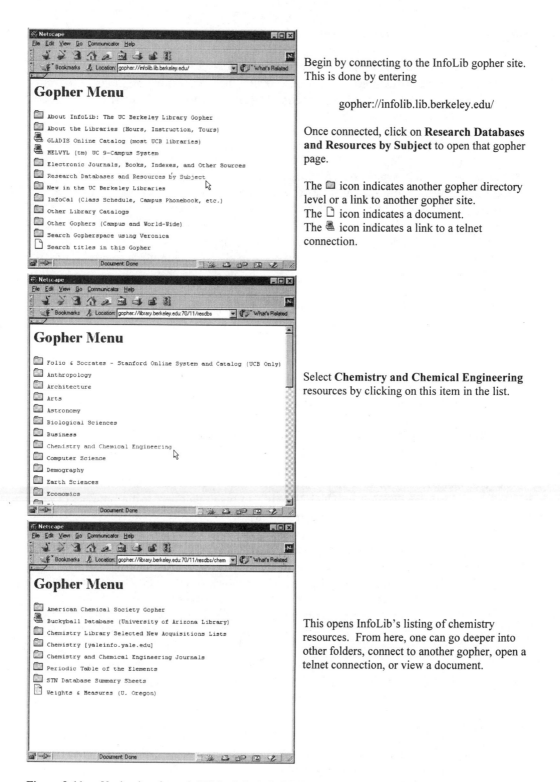

Begin by connecting to the InfoLib gopher site. This is done by entering

gopher://infolib.lib.berkeley.edu/

Once connected, click on **Research Databases and Resources by Subject** to open that gopher page.

The 🗀 icon indicates another gopher directory level or a link to another gopher site.
The 🗎 icon indicates a document.
The 🖳 icon indicates a link to a telnet connection.

Select **Chemistry and Chemical Engineering** resources by clicking on this item in the list.

This opens InfoLib's listing of chemistry resources. From here, one can go deeper into other folders, connect to another gopher, open a telnet connection, or view a document.

Figure 2-11. *Navigation through UC-Berkeley's InfoLib gopher site to find chemistry resources.*

Telnet

Text-based communications by telnet are not handled directly through your browser; however, there are Java applets to enable the opening of a telnet portal within a web page. Telnet applications are often included with the operating system (as is the case with Windows and Unix) and so do not usually need to be handled through a browser.

Other Useful Software

Aside from having software to access the Internet, it is also advisable to have supplemental programs that either prepare data for sending to an Internet site or process files once they arrive on your system. Perhaps the most important of these *helper applications* are chemical structure drawing programs, 3-D molecular structure viewers, and file decompression software.

Chemical Structure Drawing Programs

Sooner or later, every chemist needs to submit a report with drawings. While you can use a number of commercial chemical drawing programs available, there are two high-quality free programs available from the Internet. The first is *ISIS/Draw*,[40] part of the larger commercial ISIS software package for establishing chemical databases (ISIS software is published by MDL, who also publish Chime). With ISIS/Draw, one can create 2-D and 3-D line drawings, or import a 3-D structure from another source and convert it into a line drawing. ISIS/Draw, when combined with Chime, can also be used to make structure queries over the Internet, such as when searching a database or submitting a structural answer to a test question. This will be discussed in detail in the next chapter.

At some point you may wish to create your own web pages, in which case you will want to download a free copy of *ChemWeb*[41] from SoftShell/Bio-Rad.[42] This program is also a 2-D chemical structure drawing program, with similarities to ISIS/Draw, but it allows you to save your drawings as image files, which are ideal for inclusion in a web page. ChemWeb is a "lite" version of their more elaborate ChemWindows software.

Molecular Structure Viewers

From time to time, there will be advantages to viewing 3-D molecular structure files off-line, in which case a viewer would be handy. RasMol,[43] the progenitor to Chime, is still available. One drawback to using the older RasMol program is that one must designate the file format, something that is done automatically in Chime. Chime 2.0 is a much more sophisticated program than RasMol, and can be used as an off-line viewer through a browser.

---

[40] a) http://www.mdli.com/tech/isis.html#draw
    b) http://www.mdli.com/download/idraw.html
[41] http://www.softshell.bio-rad.com/FREE/ChemWeb/ChemWeb.html
[42] http://www.softshell.bio-rad.com/
[43] http://www.umass.edu/microbio/rasmol/

Another version of RasMol, "Berkeley-enhanced" RasMol,[44] offers the advantage of being able to call up multiple molecules in the same window (up to five), and uses a control panel to afford the user easier access to the command line interface. The control panel allows the user to measure bond lengths, angles, and dihedrals ("twist angles"), to rotate the molecules, and to rotate a particular bond in the molecule.

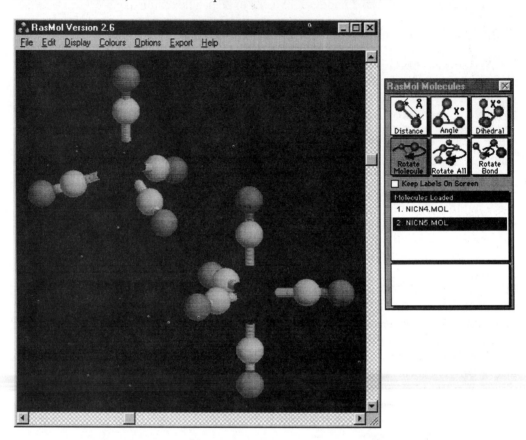

Figure 2-12. *Berkeley-enhanced RasMol provides the user with a control panel and the ability to display more than one molecule at a time.*

An off-line viewer that offers many unique and useful features is *WebLab Viewer*[45] (available as a free download from Molecular Simulations, Inc. (MSI) in a "Lite" version). WebLab Viewer is capable of using many display modes not found in Chime, including the display of repeating unit cells (Figure 2-13). WebLab Viewer is capable of reading many more file formats than Chime, and saves to a number of formats as well, including saving structures as flat 2-D images (ideal for use in printed documents) and as VRML files. WebLab Viewer is also an Internet tool; if you know the URL of a web-based molecular structure file, WebLab Viewer can retrieve it directly from the Internet without the need to use a browser.

---

[44] http://mc2.cchem.berkeley.edu/Rasmol/v2.6/
[45] http://www.msi.com/solutions/products/weblab/viewer/index.html

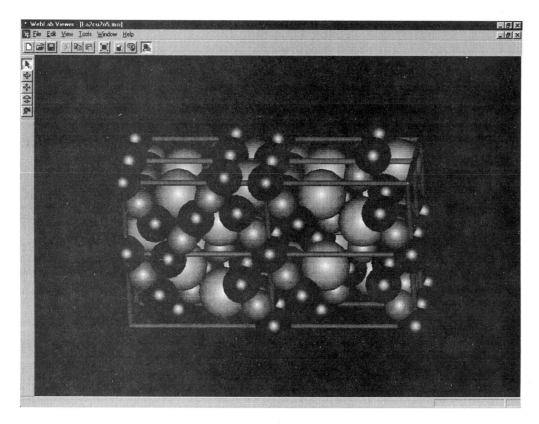

Figure 2-13. *WebLab Viewer offers display modes not found in Chime, including the display of repeating unit cells (as shown for this superconductor).*

A tool specifically designed for viewing biological molecules such as proteins and nucleic acids is *Kinemage*.[46] This helper application allows descriptive text and captions to be shown alongside the molecule and provides a number of display modes for the molecule (see Figure 2-14). A drawback to Kinemage is that it can only read its own .kin format files. Most often, proteins and nucleic acid sequences are described by what is called the Brookhaven Protein Data Bank (PDB) format;[47] therefore, a program called *PreKin* that converts .pdb files into .kin files is usually enclosed with Kinemage. Both of these programs may be downloaded from the Protein Science web site, which also provides instructions on using Kinemage in conjunction with your browser (for more modern browsers, this is a very simple task: the first time you select a link to a Kinemage file, the browser asks what application you'd like to use to view the file type; you respond with the location of your Kinemage program, which sets the browser for future links to this same file type). An *Exploring Molecular Structure* software package is enclosed with most Prentice Hall biochemistry textbooks and contains the Kinemage program and a number of molecular structure files.

---

[46] http://prosci.org/Kinemage/
[47] http://www.pdb.bnl.gov/

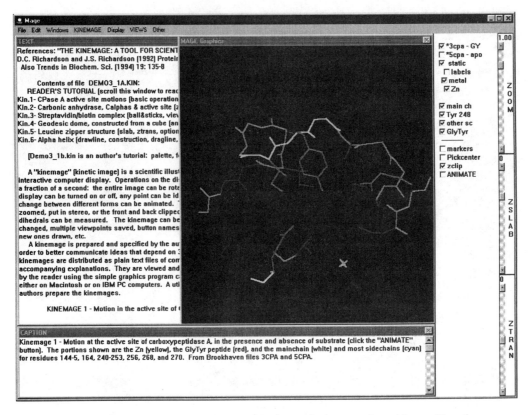

Figure 2-14. *The Kinemage interface allows captioning and other text alongside graphics of biomolecules.*

A molecular structure file can come in various formats, but almost all are text files that give spatial coordinates of the atoms in a molecule, and usually information on how the atoms are bonded to each other. Although most viewers are capable of handling a range of file formats, there are far more formats in existence than any one viewer can possibly be designed to accept. A program that is useful for interconverting between molecular structure file formats is *Babel*.[48] Babel is available for all major platforms. A list of the various file types that Babel can read and write is shown in Table 2-1.

File Compression/Decompression Software

One last type of software you will need at some point is one or more applications for file compression and decompression. Almost every program file transmitted over the Internet has been compressed in some way to save transfer time, and it must be decompressed before you can install and run it. The software repositories will likely offer these sorts of file manipulation programs. Beyond this, a popular vendor for this sort of software is *Aladdin Systems*,[49] which offers compression/decompression programs as shareware and commercialware for both Windows and Macintosh platforms.

---

[48] http://mercury.aichem.arizona.edu/babel.html
[49] http://www.aladdinsys.com/

Babel input file types	Babel output file types
Alchemy	Alchemy
AMBER PREP	Ball and Stick
Ball and Stick	Cacao Cartesian
Biosym .CAR	CAChe MolStruct
Boogie	Chem3D Cartesian 1
Cacao Cartesian	Chem3D Cartesian 2
Cambridge CADPAC	ChemDraw Conn. Table
CHARMm	CSD CSSR
Chem3D Cartesian 1	Gamess Input
Chem3D Cartesian 2	Gaussian Cartesian
CSD CSSR	Gaussian Z-matrix
CSD FDAT	Hyperchem HIN
CSD GSTAT	IDATM
Free Form Fractional	Mac Molecule
GAMESS Output	Macromodel
Gaussian Z-Matrix	Micro World
Gaussian Output	MM2 Input
Hyperchem HIN	MM2 Ouput
Mac Molecule	MM3
Macromodel	MMADS
Micro World	MDL Molfile
Molgen	Molgen
MM2 Input	Mopac Cartesian
MM2 Ouput	Mopac Internal
MM3	PDB
MMADS	Report
MDL MOLfile	Spartan
MOLIN	Sybyl Mol
Mopac Cartesian	Sybyl Mol2
Mopac Internal	XYZ
Mopac Output	
PDB	
Quanta	
ShelX	
Spartan	
Sybyl Mol	
Sybyl Mol2	
XYZ	

Table 2-1. *A list of the molecular structure file formats to which Babel can read and write.*

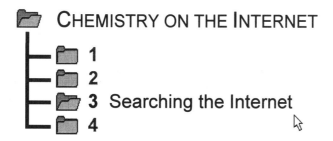

Searching the Internet

Now that you've made a connection to the Internet, and have the tools necessary to find your way around, the question remains, just how *do* you find your way around those 30 million servers waiting to feed you information? Naturally, you can locate new sites through the links in pages you already visit, but that can be a very limited way of finding new information. Fortunately, there are well over a hundred different search tools[50] in existence, ranging in coverage from the very general to the very specific.

General Search Tools

Searching the World Wide Web

In searching the WWW, there are two approaches available: *web directories* and *search engines*. A web directory acts as a table of contents to the WWW (and often much of the rest of the Internet). A prime example of a web directory is *Yahoo*.[51] This resource, started by two graduate students at Stanford University as a list of favorite sites (a *hotlist*), has grown to be a very large and well-organized guide to Internet resources. Yahoo accumulates its listings largely from user submissions. An added advantage to Yahoo is that it allows you to pass a search request along to a variety of other search tools, thus providing a search *gateway*. A great many other web directories have cropped up in the interim, including *Excite*[52] and *Snap*,[53] among others. Competition between web directories has become so great that many of them are branching out to offer other services, such as web-based e-mail, local news, weather and media listings, and even auctions.

A search engine is a somewhat different tool. Its function is to allow you to do keyword searches through Internet listings that have usually been gathered by *robots* and *spiders* (automated software programs that wander the web and accumulate information on the titles and contents of web pages). The search feature you use then scans this database of information and returns the sites from which they were obtained. Popular search engines include Alta Vista[54] and Lycos.[55] Many search engines and web directories work together; for example, a search of Yahoo also uses the search engine Alta Vista, and lists those results after the Yahoo listings.

---

[50] http://search.cnet.com/Alpha/1,6,0,0200.html
[51] http://www.yahoo.com/
[52] http://www.excite.com/
[53] http://www.snap.com/
[54] http://www.altavista.com/
[55] http://www.lycos.com/

Because of the sheer number of search options available today, many sites have cropped up that offer nodes for searching by multiple methods (these are sometimes referred to as *searchbots*). *Search.com*[56] is one such example, offering access to eleven search engines simultaneously with its Express Search option. Netscape[57] also offers a page that will access multiple search engines; this can be summoned easily with the Netscape browser by clicking on the flashlight-shaped icon in the Navigation Toolbar.

A useful cousin to the web directory is the resource listing, or *meta-index*. Many individuals put up these web pages with listings of sites on a particular subject. The specificity of these listings is often more in-depth than a more general web directory provides. The *Chemistry on the Internet*[58] resource site is one such listing for chemistry; others can be found in most web directories, such as the listing at Yahoo[59]).

Searching Usenet News and Mailing Lists

Usenet postings could conceivably be very difficult to search using a newsreader; this is complicated by the fact that articles only remain posted for a few days. Fortunately, the *DejaNews Research Service*[60] and *reference.com*[61] web sites (Figure 3-1) archive Usenet postings, and their article databases are searchable by keyword. Reference.com also archives many of the Internet mailing lists.

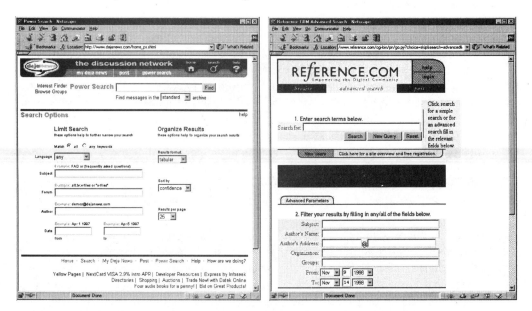

Figure 3-1. *The DejaNews Research Service and reference.com are two tools for searching archives of Usenet News and mailing list discussion postings.*

---

[56] http://search.cnet.com/
[57] http://search.netscape.com/
[58] http://acad.tnstate.edu/~chemnet/
[59] http://dir.yahoo.com/Science/Chemistry/Web_Directories/
[60] http://www.dejanews.com/
[61] http://reference.com/

Searching FTP Sites - *Archie*

If you're looking for something that might be found on an FTP site, the tool of choice is *Archie* (derived from the word "archive" and based, as are the tools that follow, on the Archie comics series[62]). There are several Archie servers around the world, which can be accessed via the WWW using *ArchiePlex*.[63] A limitation of Archie is that you need to know the name of the file you're seeking, or at least part of it. Other Archie web gateways[64] are available, as well as specific Internet client tools for using Archie.[65]

If you're searching for a program to perform a specific task, you might consider doing a search for a web page on that subject, as most page authors include links to other sites that might have what you're looking for. This type of search can often be much more rewarding than an Archie search. Some places to start are listed below:

TUCOWS	The Ultimate Collection of Winsock Software (TUCOWS) is one of the largest cross-platform repositories of Internet software on the web. Software for Windows 3.1, 95, 98, NT and Macintosh is available here, as well as several Java applets. TUCOWS has many mirror sites worldwide, listed at their home site,[66] for easy download of files, and files can also be obtained directly from their central repository.[67]
Winfiles.com	Another large repository that is devoted to Windows-based software is Winfiles.com[68] (formerly Windows95.com).
Meta-indices	Resource listings on a particular subject that are created by individuals often contain links to software. One specific example for chemistry is the *Chemistry on the Internet*[69] FTP site[70] maintained by *The Chemical Educator* journal.[71]
Corporate FTP sites	Software can, of course, usually be obtained directly by downloading it from a corporate web site. URL addresses of many software publishers have been listed in footnotes throughout this book.

---

[62] http://www.archiecomics.com/
[63] a) http://pubweb.nexor.co.uk/public/archie/servers.html
b) http://cuiwww.unige.ch/archieplexform.html
[64] http://dir.yahoo.com/Computers_and_Internet/Internet/FTP_Sites/Searching/Archie/
[65] a) http://www.fpware.demon.nl/
b) http://ds.dial.pipex.com/d.woakes/
c) http://www.stairways.com/anarchie/
[66] http://www.tucows.com/
[67] http://idirect.tucows.com/
[68] http://www.winfiles.com/
[69] http://acad.tnstate.edu/~chemnet/
[70] http://acad.tnstate.edu/~chemnet/ftp.html
[71] http://www3.springer-ny.com/chedr/

Searching Gopher Sites - *Veronica* and *Jughead*

The tool for searching gopher is *Veronica*[72] (an acronym for Very Easy Rodent-Oriented Net-wide Index to Computerized Archives). Most gopher sites have links to either their own Veronica or one at another site. Veronica searches all gopher sites for title keywords. A web-based interface to Veronica can be found at the Galaxy-Einet gopher information site.[73] A related tool, *Jughead*,[74] searches the titles in a single, local gopher site. This is particularly useful for finding a specific document within a large site.

Just as web pages have largely supplanted gopher pages since the widespread introduction of HTML, web search tools have replaced Veronica and Jughead. Most web directories and search engines will list gopher sites alongside web sites, as well as telnet connections.

Chemistry-Related Searches

Aside from the general search methods, there are a variety of subject-specific search resources on the Internet, especially for science and technology. One of the largest information databases in existence is maintained by the *Chemical Abstracts Service* (CAS),[75] and this commercial resource is searchable over the web. The CAS web site describes their various information products, as well as specialized search tools suited to use on a web page.

Chemical patent information is another major search area, and CAS offers *Chemical Patents Plus*[76] to meet this need. Another source of patent information is the *IBM Intellectual Property Network* (formerly the IBM Patent Server),[77] which offers free searching and downloading of document images for patents dating back to 1971.

Searching by Structure

When searching for information on a chemical compound, it would be ideal to somehow be able to relate the structure to the computer and search on that basis. With our modern tools, we can do such a search by one of two methods: either by drawing the structure separately and transferring it to a web page, or by using a plug-in or Java applet to draw the structure on the web page and enter it directly.

An example of the first is seen in the use of MDL's ISIS/Draw and Chemscape Chime in a search of the Hazardous Substances Databank at NIH's Specialized Information Services (SIS) Chemical Structure Searching web site (Figure 3-2).[78] A structure draw in ISIS/Draw is copied in the MDL Molfile format, and then pasted into a Chime window, which then passes it to a Chemscape server for processing. Since ISIS/Draw can also import molfiles, molecules from other sources can be used for searches as well.

---

[72] http://dir.yahoo.com/Computers_and_Internet/Internet/Gopher/Searching/Veronica/
[73] http://galaxy.einet.net/gopher/gopher.html
[74] gopher://gopher.utah.edu/11/Search%20menu%20titles%20using%20jughead/Search%20other%20 institutions%20using%20jughead
[75] http://www.cas.org/
[76] http://casweb.cas.org/chempatplus/
[77] http://www.patents.ibm.com/
[78] http://chem.sis.nlm.nih.gov/

Figure 3-2. *Structure Searching with ISIS/Draw and Chime*
Step One: *Configure ISIS/Draw to Copy Structures as MolFiles*

Before you can use ISIS/Draw to create structures for use in Chime-based database searches, you must configure it to allow copying and pasting in the MolFile format. (Once this is set, it remains set for all future sessions unless you change it otherwise.)

To begin, in the ISIS/Draw Options menu, select Settings.

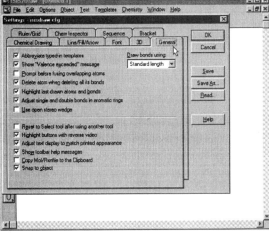

Select the General options card.

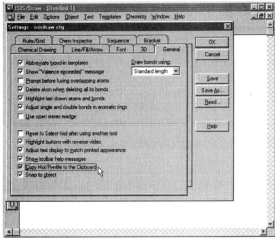

Select the option Copy Mol/Rxnfile to the Clipboard.

Click OK to finish. Now you will be able to transfer structures from ISIS/Draw to Chime.

Figure 3-2. (*continued*)
Step Two: *Create and Copy a Structure in ISIS/Draw*

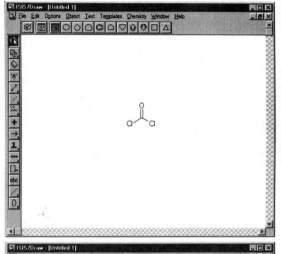

Use ISIS/Draw to sketch the molecule or fragment that you wish to search. (In this case, a phosgene molecule has been drawn.)

Keep in mind that MDL molfiles from other sources can also be used by importing them into ISIS/Draw.

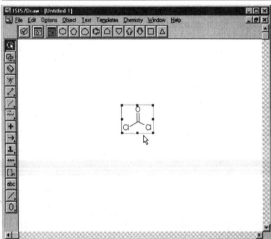

Select the molecule.
This can be done using the select button (at the top of the vertical toolbar) and the cursor, or, if the molecule is the only item in the window, use <u>Select All</u> from the <u>Edit</u> menu.

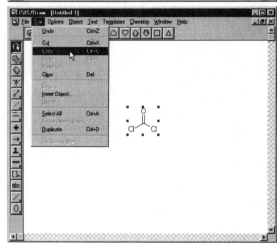

Copy the molecule.
This can be done by clicking the right mouse button over the structure and selecting <u>Copy</u>, or by selecting <u>Copy</u> from the <u>Edit</u> menu.
From here, go to the web database you wish to search.

Figure 3-2. (*continued*)
Step Three: *Enter the Structure into the Chime Window of the Database Search Engine*

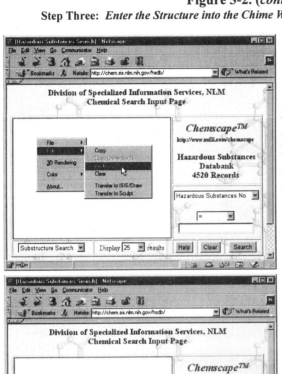

For this example, we have opened the Hazardous Substances Databank through the NIH's Specialized Information Services (SIS) Chemical Structure Searching web site at http://chem.sis.nlm.nih.gov/.

Using the right mouse button to click on the Chime window and open the pop-up menu, select Paste from the Edit menu.

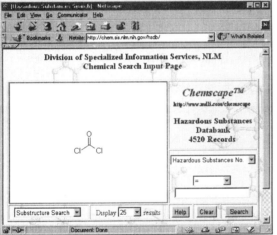

The structure now appears in the Chime window. The various other parameters of the search may be modified if necessary, and the Search button clicked to initiate the process.

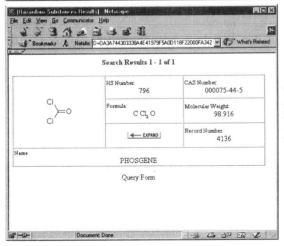

The results of the search are displayed as a new page. The structure is shown in a new Chime window, and as such, it can be switched to 3-D Rendering using the pop-up Chime menu.

39

There are two methods of drawing a structure in a web page. The first uses the CambridgeSoft[79] (CS) product *ChemOffice*, which is a commercial drawing and visualization package (a combination of their products *ChemDraw*, for 2-D renderings, and *Chem3D*). CS offers free "Net" versions of a Chem3D plug-in[80] that allows visualization of molecules (much like Chime) and also a ChemDraw plug-in[81] for the direct drawing of structures into the plug-in window with the aid of a ChemDraw toolbar. Both plug-ins can be obtained in a combined form as a ChemOffice Net plug-in. (It should be noted that one only needs the ChemDraw plug-in for search applications, and that the Chem3D plug-in may cause compatibility problems with a Chime plug-in if both are loaded.) The example shown in Figure 3-3 is the use of the ChemDraw Net plug-in to search *CS ChemFinder*[82] for information on a chemical compound. *ChemFinder* is a tool that will help you find information about a particular compound. It allows you to search for a chemical based on name, formula, molecular weight, Chemical Abstracts Service (CAS) registry number, or structural features. Like a web directory, *ChemFinder* relies on submissions from users.

The second method uses a commercial Java applet called *ChemSymphony*.[83] This applet is used mainly for visualization of structures, but can also be configured to permit drawing of structures in a separate window, which then may be passed to a web page for use in a search engine. An example of this is a search of the NCI-3D database through the *WWW Chemicals* web site (Figure 3-4).[84]

In addition to using these methods for on-line searches, they are also a way of answering test questions with pictorial answers. The same algorithm that processes a search for matching records in a database can also compare a submitted structure to the possible correct answers for a question. This technology is likely to be implemented on many education web sites in the very near future.

---

[79] http://www.camsoft.com/
[80] http://www.camsoft.com/plugins/chem3D.html
[81] http://www.camsoft.com/plugins/chemdraw.html
[82] http://chemfinder.camsoft.com/
[83] http://www.chemsymphony.com/
[84] http://www.chem.com/structures/

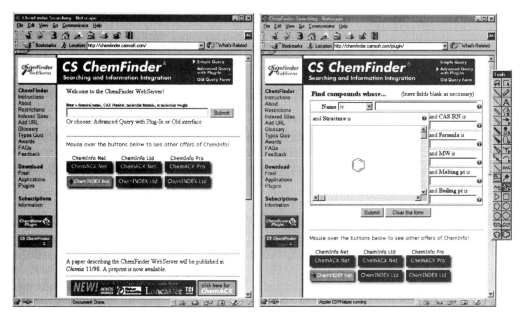

Figure 3-3. *Searching CS ChemFinder using either the text interface (left) or the structure drawing interface (right) with the CS ChemOffice plug-in and ChemDraw toolbar.*

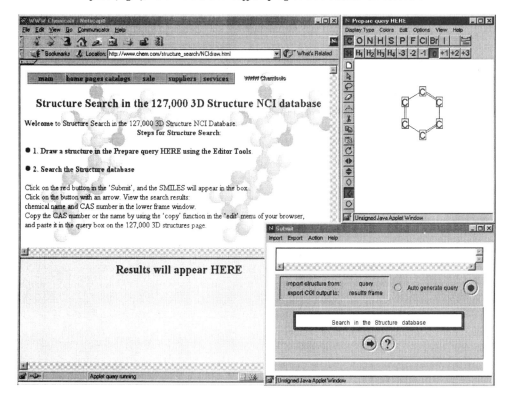

Figure 3-4. *Searching the NCI-3D database with the ChemSymphony applet. A structure is drawn in one applet window (upper right), then the second window (lower right) passes the information to the web page (left) for processing.*

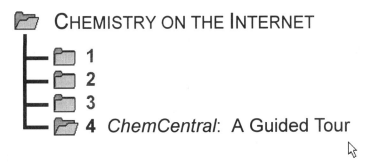

CHEMISTRY ON THE INTERNET

- 1
- 2
- 3
- **4** *ChemCentral*: A Guided Tour

With web directories and other search tools, you can find just about anything that interests you on the Internet. In this last chapter, your attention will be directed to those resources offered by the Prentice Hall chemistry web site, *ChemCentral*. As has been mentioned on occasion throughout this book, *ChemCentral* offers many unique functions to aid you in your study of chemistry.

The home page for this web site was presented to you back in Figure 2-1. Let's begin our tour by selecting the General Chemistry section of the main menu, either by clicking on the appropriate section of the map, or by selecting the button in the left-hand column. This opens a selection of textbooks (Figure 4-1), from which you may select the book that you use in your course. In this example, let's go into the area for *Chemistry: The Central Science* by Brown, LeMay and Bursten.

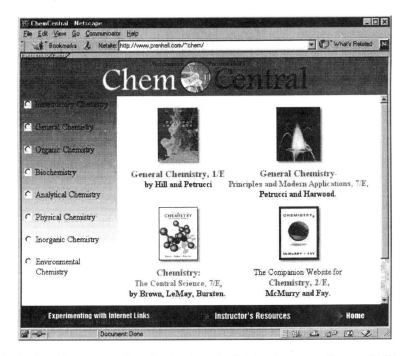

Figure 4-1. *Each subject area of* **ChemCentral** *contains links to the various Prentice Hall textbooks that represent that area.*

In each textbook area, you will find a toolbar on the left-hand side of the main screen that will direct you to the various resources for that book. Let's take a look the options one might find in a typical toolbar.

chapter 1 — This is a welcoming area that provides you with an outline review of the concepts introduced in the text.

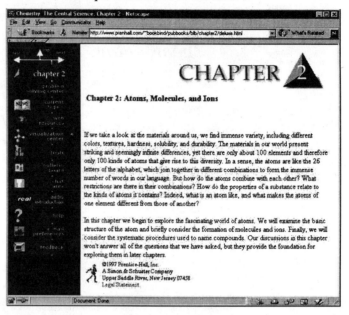

problem solving center — How well did you understand the material in the chapter? Are you ready for the next exam? This is an area in which you can try your hand at practice exercises and exams. Pre-med students will find questions related to the MCAT exam here.

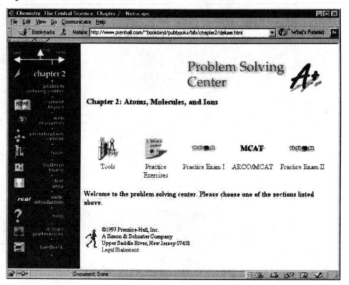

 This area will show you a practical "real world" application of the concepts discussed in the text, as drawn from contemporary news features, and allow you to practice your understanding of the concepts with short essay questions. Visual aids and editorial comments clarify the real-world application and connect it to your reading in the text.

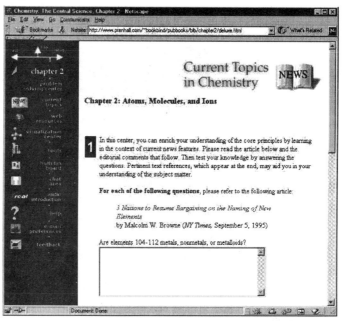

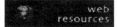 Sites on the Internet that offer information relevant to the chapter, or to the book in general, are listed here. This listing is continually updated to ensure that all links remain current. The links also indicate the need for special plug-ins before viewing.

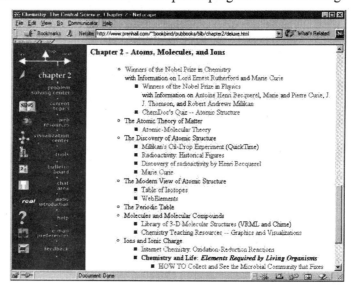

 This is the place to go to "reach out and touch" a molecule. Use your Chime plug-in to see molecules from each chapter presented in 3-D. While most of the molecules presented in this section can be viewed with Chime 1.0, it is strongly advisable to get Chime 2.0 to be ready for new Visualization Center features that utilizes the newer version's more advanced features.

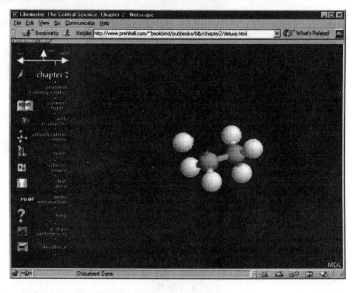

 This section contains periodic tables and other chemical information that will be useful in studying the chapter and performing the exercises.

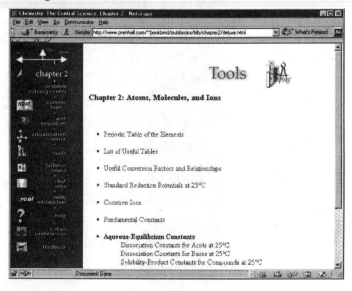

 This is a Usenet newsgroup established for posting messages to discuss of any aspect of the chapter with students and teachers elsewhere. This can be a good place to come for homework help.

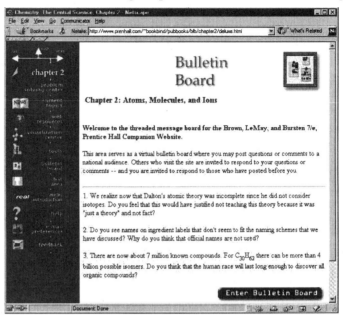

 This area offers live chat with other students using the same textbook. This can be particularly effective for study group meetings pre-arranged by e-mail. This feature uses the *iChat* plug-in.[85]

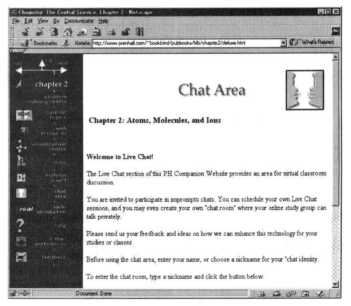

---

[85] http://www.acuity.com/ichat/download/client.html

 This is a streaming audio introduction to the subject matter contained in the chapter. It requires you to have the Real Player software program (described in Chapter Two).

 This is an on-line help guide to all aspects of the *ChemCentral* site. It will give you any necessary supplemental instruction in the use of the web site.

 After working the exercises on this web site, you can make arrangements to have the results e-mailed to yourself, your instructor, or someone else by making the appropriate selections in the area.

 How do you feel about this site? Are the exercises at the appropriate level of difficulty? Are there any features you would like to see added? Is there anything else you would change? This is the place to leave such comments.

Other textbooks in the *ChemCentral* web site have most if not all of these features built into their areas, with more features being built all the time. As such, *ChemCentral* is a site that can serve you throughout your college years as a supplement to your Prentice Hall textbooks. You are indeed most fortunate to be studying chemistry in this day and age, where such advanced computer tools are at your disposal!

Glossary

This is a short list of the terms mentioned in this book, as well as some that you might encounter as you explore the Internet. A thorough (and multilingual) glossary of Internet terms is *NetGlos*, at `http://wwli.com/translation/netglos/netglos.html`. A more general glossary of computer terminology is the *Free On-line Dictionary of Computing* at `http://www.instantweb.com/~foldoc/`.

ActiveX	a method for adding specialized programs to the Microsoft *Internet Explorer web browser*
Archie	a tool for searching *FTP* sites
article	a message *posted* to *Usenet News* or a *mailing list*
ascii	a term for a text-only file (stands for American Standard Code for Information Interchange)
attachment	a *binary* file sent with an *e-mail* message or a *Usenet* posting
Babel	a program for converting between *molecular structure file* types
background	a pattern or color appearing behind the text and images of a *web page*, like wallpaper
BBS	Bulletin Board System - a place for discussing topics by posting and replying to messages
binary	a non-text file, such as a program, image, sound file, or animation
bookmark	a *web page* that you plan to revisit, as recorded by the *web browser*
bit	a *binary* digit; a unit of binary information (1 and 0); there are eight bits in a *byte*
bps	Bits Per Second - a unit of modem speed; modems generally operate at thousands of bps (Kbps) or millions of bps (Mbps)
browser	a software program for presenting *Internet* (and especially *World Wide Web*) information, usually in a graphical fashion
byte	a unit of size for computer files; disks are usually measured in millions of bytes, or megabytes (MB)
cache	a temporary storage space for computer data, either in memory or on disk
card	a supplemental circuit board for a computer, as in an *Ethernet* card, or a *modem* card
CGI script	Common Gateway Interface - a type of *program* that a *web page* uses to perform a specialized function

Chime	a plug-in for viewing and manipulating *molecular structures*
client	a computer or *program* that reads information from the *Internet*
commercialware	*software* sold for profit
CSLIP	Compressed Serial Line Internet Protocol - one of three methods used by a *TCP stack* to communicate with the *Internet* over a phone line and allow the use of *client software*; see also *SLIP* and *PPP*
demoware	*software* distributed for demonstration purposes, with a limited set of capabilities compared to the full version, or a limited length of time before it ceases to function
domain	a group of computers with a common purpose, such as governmental, military, educational, commercial, or a particular foreign country; the two or three letter code at the end of a *domain name*
domain name	an identifying term for a computer *host*, consisting of two or more words separated by periods, the last of which is the *domain*
DOS	Disk Operating System - the set of *programs* that control the function of IBM Personal Computers (PC) and related computer types (PC clones)
download	the process of obtaining a file from a remote computer; see also *upload*
e-mail	Electronic Mail - a method for sending text messages and *binary* attachments to a person or group of people over the *Internet*
embed	to place one type of document within another type of document; for example, a *molecular structure* file is embedded in a *hypertext* document for viewing with *Chime*
Ethernet	the term for the mechanical aspects of connecting a *Local Area Network*; for example, a computer connected to a LAN will use an Ethernet *card*
FAQ	Frequently Asked Questions - a list, used most often in *Usenet newsgroups* and other discussion media, designed to eliminate the need for *posters* (especially new posters) to ask routine questions about the topic at hand
floating menu	a menu that appears anywhere on screen when a certain mouse button is pressed at that point
freeware	*software* that is distributed without the need for a registration fee or other cost
FTP	File Transfer Protocol - the method for transferring text and *binary* files between *Internet* nodes
Gopher	a method for presenting text-only lists with link capabilities

GUI	Graphical User Interface - a user interface that uses icons (small pictures) and other graphics to facilitate access to the command structure of the computer
home page	a *web page* that acts as the first or welcoming page to a set of other pages in a *web site*; a page that provides information about a person or specific subject
hotlist	a list of *bookmarks* or other favorite sites
HTML	HyperText Mark-up Language - the marking of a text document to provide formatting instructions to a *web browser*; see also *tags*
HTTP	HyperText Transfer Protocol - the protocol for transferring multimedia and *hyperlinked WWW* documents over the *Internet*
hyperlink	a reference in a *hypertext* document that, when activated (usually by a mouse click), will connect the *web browser* to the referred *web page*
hypertext	text that contains *hyperlinks* to other documents
image map	also known as a clickable map; an image that contains built-in *hyperlinks*, such that clicking on a part of the picture will take you to a certain web page
in-line	an image or other non-text object inserted into a body of text in a formatted manner
Internet	a global collection of interconnected computer networks
Internet Explorer	a *web browser* from Microsoft
Intranet	another term for a *Local Area Network* (LAN)
IP address	a set of four numbers, each between 0 and 256, and separated by decimal points, that define a unique Internet host computer; see also *domain name*
ISDN	Integrated Services Digital Network - a specialized *modem* and telephone line that sends data faster than conventional analog systems by using signal digitization
ISP	Internet Service Provider - a company that allows you to dial in with your computer and modem, and will connect you to the *Internet*, ideally providing you with *e-mail* service and a *Usenet* newsfeed
Java	a method for sending specialized programs to a *web browser*
JavaScript	a language set of scripted instructions that enhance the performance of *HTML* and create special effects on a *web page*
Jughead	a tool for searching a local *gopher* site
Kinemage	a *helper* application for viewing biological molecules

LAN	Local Area Network - a set of interconnected computers that may or may not be hooked into the Internet
listserver	a computer for receiving and redistributing *e-mail* messages in a *mailing list*
log	a record of a transaction of computer data
log-in	a procedure for gaining access to a computer system or account by entering a user name and password
Macintosh	a *GUI* computer *platform* produced by Apple Computer, Inc.
mailing list	a method for communicating amongst a group of people by *e-mail* on a particular topic; mail is sent to a *listserver*, which then redistributes the message to all list subscribers
mainframe	a centralized computer that accepts multiple simultaneous users through a time-sharing mechanism
meta-list	a synonym of *hotlist*
mirror	a *server* that duplicates information found on another *server*
modem	a device for communicating with a computer over a telephone line (short for modulator-demodulator)
moderated	a *newsgroup*, *mailing list*, or *BBS* forum that is overseen by someone for content
molecular structure file	a file, usually in a text format, that contains 3-D coordinate locations of atoms in a molecule, and usually some information describing how the atoms are connected
Mosaic	one of the first *web browsers*
multimedia	the combination of any or all of the media forms such as text, pictures, animations, sound, etc.
Netscape Navigator	a *web browser* from Netscape Communications
network	any group of two or more computers which are interconnected such that they can share data
newsfeed	a full set of *Usenet postings* sent from a *news server*
newsgroup	a section of *Usenet News* devoted to a specific topic
newsreader	a *program* for reading and organizing *Usenet News* articles
news server	a computer that provides *Usenet News*
NNTP	Network News Transfer Protocol - the process of transferring *Usenet News*; the address of a *news server*
node	any computer connected to the *Internet*, including both *clients* and *servers*

platform	a computer type or operating system, e.g. *Macintosh*, *Windows*, *DOS*, *Unix*, etc.
post	*n.* an *article* sent to a *Usenet newsgroup* or a *mailing list* *v.* to send an *article*
PPP	Point-to-Point Protocol - the best of three methods used by a *TCP stack* to communicate with the *Internet* over a phone line and allow the use of *client software*; PPP allows error detection on the communication line, as well as automatic *IP address* assignment; see also *SLIP* and *CSLIP*
processor	the main circuit chip computer, the one which, more than any other, defines the speed of your computer
program	also called *software*; a set of instructions used by the computer to tell it how to perform a certain function or handle data
RAM	Random Access Memory - the circuit chips in your computer responsible for holding programs and data in use; more memory generally allows faster operation and permits the use of more sophisticated *software*
RasMol	a *helper* application for viewing and manipulating images of *molecules*
readme	a text file that provides information, for example, about the contents of an *FTP* area
remote log-in	another term for *telnet*
server	a computer *host* that provides information to the *Internet*, such as a *web server* or *news server*
shareware	*software* distributed in exchange for a voluntary registration fee
SLIP	Serial Line Internet Protocol - one of three methods used by a *TCP stack* to communicate with the *Internet* over a phone line and allow the use of *client software*; see also *CSLIP* and *PPP*
software	a set of instructions, or *program*, used by the computer to perform a certain function
T-1 line **T-3 line**	the backbones of the *Internet*, the main types of communication lines that interconnect local networks; a T-1 carries data at 1.54 *Mbps*, and a T-3 carries it at 44.3 Mbps
tags	the marks put into an *HTML* document to provide formatting instruction for a *web browser*, identified by their < > brackets
telnet	a method for *logging in* to a remote computer system; also called *remote log-in*
thread	a series of *Usenet* or *mailing list* articles on the same subject, forming the basis of a discussion

Unix	a type of computer commonly used for high-end *Internet* work, and designed to *serve* multiple simultaneous users
upload	a process for sending information from a local computer to a remote computer; see also *download*
URL	Universal Resource Locator - the instruction line that gives direction to a *web browser*, and uniquely identifies each *web page*
Usenet	an area of the *Internet* devoted to the *posting* of *articles* on thousands of different topics
Veronica	Very Easy Rodent-Oriented Net-wide Index to Computerized Archives - a method for searching multiple *gopher* sites, or *gopherspace*
web page	another term for a *hypertext* document
web site	a set of *web pages*, or another term for a *web server*
web server	a computer that provides *hypertext* information
whiteboard	a *software* tool that allows users to draw (using a mouse or other pointing device) an image on-screen that is simultaneously transmitted to other Internet users
Windows	a *GUI* interface for *DOS* developed by Microsoft Corporation
WWW	World Wide Web - the subset of the *Internet* devoted to transferring formatted multimedia *hypertext* documents
WYSIWYG	What You See Is What You Get (pronounced as *whizzy wig*) - a term referring to the ability of some editing programs or other document preparation software to display a document on screen in exactly the same format as it will be printed or otherwise published

Index

A

ActiveX 14, 22-23
    security certificates 22

B

bookmarks 11-13
bulletin board systems (BBS) 4

C

chat 4, 6, 47
clients 1, 9

D

domains 9-10
    country code 9
    domain name system (DNS) 9-10

E

Electronic Mail (E-Mail) 4-5, 13, 25, 33, 47
    mailing lists 4, 34
ethernet 7

F

File Transfer Protocol (FTP) 4-5, 10-11, 25, 35
finger 6
Frequently Asked Questions (FAQ) 4

G

gopher 5, 10-11, 25-26, 36
    gopher sites
        Galaxy-Einet 36
        University of California at Berkeley's *InfoLib* ....
        25-26
graphical user interfaces (GUI) 5, 7, 9

H

hosts 1-3, 9-10
hypertext 5-6, 10
hypertext markup language (HTML) 5, 23, 36
    forms 23-24

I

image map 11
Internet
    growth 2
    history 1

Internet protocol (IP) address 9-10
    dynamic 9
    static 9
Internet service provider (ISP) 8-9, 11, 25

J

Java 14, 19-23, 27, 35-36, 40
    applets 19
        ChemSymphony 20, 40-41
        FormulaOne 21
        Molecule Viewer 20
    spreadsheets 21
Sun's *Java Development Kit* (JDK) 21
virtual machines 19

L

listservers 4
local area network (LAN) 7

M

modem 7-9

N

nodes 1-2, 9, 34

P

plug-ins 14, 22, 40, 45
protocols 3, 9-10, 25
protocols, dial-up communications
    compressed SLIP (CSLIP) 9
    point-to-point protocol (PPP) 9
    serial line Internet protocol (SLIP) 9
push technology 6
Push Technology 6

R

remote log-in .... See telnet
router 9

S

scripts 14, 23
    client-side 23
    JavaScript 23
    JScript 23
    server-side 23
        PERL 23
    VBScript 23

searching
   chemistry-related 36
      CambridgeSoft
         ChemFinder 40-41
      Chemical Abstracts Service (CAS).................. 36
         Chemical Patents Plus 36
      IBM Intellectual Property Network 36
      National Cancer Institute NCI-3D database ... 40
      WWW Chemicals 40
   FTP sites 35
      Archie 35
         web gateways
            ArchiePlex 35
   gopher 36
      Jughead 36
      Veronica 36
   mailing lists
      reference.com 34
   Usenet News 34
      DejaNews Research Service 34
      reference.com 34
   World Wide Web (WWW) 33
      meta-indices 34-35
         Chemistry on the Internet 34-35
      search engines... 33
         Alta Vista 33
         robots and spiders 33
      searchbots 34
         Netscape 34
         search.com 34
      web directories 33
         Excite 33
         Snap 33
         Yahoo 33
servers 1-3, 6, 9, 33, 35
software
   Babel 30-31
   chemical structure drawing programs 27
      ChemOffice 40
         Chem3D 40
         ChemDraw 40
      ISIS/Draw 27, 36-38
   client programs 25
   compression and decompression programs 30
   helper applications 27
   molecular structure viewers 27
      ChemWeb 27
      Kinemage 29-30
         PreKin converter program 29
      RasMol 14, 27-28, 30
      WebLab Viewer 28-29
   plug-ins
      ChemOffice Net 40-41
         Chem3D Net 40
         ChemDraw Net 40-41
      FormulaOne 21
      iChat 47
      Macromedia Shockwave 14-17
      Chemscape Chime ....
         .... 14-16, 18, 20, 27-29, 36-40, 46
      QuickTime 14, 16
      QuickTime Virtual Reality (QTVR) 16

   Real Player 14, 18
   virtual reality modeling language (VRML)
      Cosmo Player 18
Prentice Hall's Exploring Molecular Structure 29
repositories
   Chemistry on the Internet 35
   TUCOWS 35
   winfiles.com 35
web browsers 11-12, 14, 19, 25
   Microsoft Internet Explorer 11, 14
   Mosaic 11
   Netscape Navigator 11, 14
Windows Media Player 18

T

telnet 3, 5, 11, 26-27, 36

U

uniform resource locator (URL) 10-11, 25, 28, 35
Usenet News 4, 25, 34
   DejaNews Research Service 25, 34
   My DejaNews 25

V

virtual reality modeling language (VRML) ...14, 18-19
   atomic orbitals 18

W

World Wide Web (WWW) 5, 7, 11, 25, 33, 35
   hotlists 33
   web sites
      Aladdin Systems) 30
      Apple 16
      BrowserWatch 14
      CambridgeSoft 40
         ChemFinder 40-41
      Chemical Abstracts Service (CAS) 36
         Chemical Patents Plus 36
      Chemistry on the Internet 34-35
      DejaNews Research Service 25, 34
      developer.com 20
      Explore Science 16
      Free On-line Dictionary of Computing 49
      Gamelan 20
      IBM Intellectual Property Network 36
      InterNIC 9
      Macromedia 16
      MDL (Molecular Design Limited) 14-15, 27
      Molecular Models 16
      MSI (Molecular Simulations, Inc.) 28
      National Cancer Institute NCI-3D database ....
         .... 16, 40
      NetGlos 49
      Netscape Communications 11
      NIH Specialized Information Services (SIS)
         Hazardous Substances Databank 36-39

Prentice Hall's *ChemCentral* 10-12, 16, 18, 20, 43, 48
Protein Science) 29
Real 18
reference.com 34
Science Daily 25
SoftShell/Bio-Rad 27
Structure and Isomerism of Coordination Compounds 14-15
Sun Microsystems 19
The Chemical Educator 35
TUCOWS 14, 35
Valence Shell Electron Pair Repulsion (VSEPR) (Purdue Univ.) 14-16
Valence Shell Electron Pair Repulsion (VSEPR) (Univ. of Sheffield) 15-16
Visual Components 21
VRML in Chemistry 18
winfiles.com 35
WWW Chemicals 40